Offshore Mechatronics Systems Engineering

T0133820

Editor

Hamid Reza Karimi

Department of Mechanical Engineering
Politecnico di Milano, Milan, Italy

CRC Press

Taylor & Francis Group
Boca Raton London New York

CRC Press is an imprint of the
Taylor & Francis Group, an **informa** business

A SCIENCE PUBLISHERS BOOK

CRC Press
Taylor & Francis Group
6000 Broken Sound Parkway NW, Suite 300
Boca Raton, FL 33487-2742

First issued in paperback 2021

© 2018 by Taylor & Francis Group, LLC
CRC Press is an imprint of Taylor & Francis Group, an Informa business

No claim to original U.S. Government works

Version Date: 20180331

ISBN-13: 978-0-367-78107-1 (pbk)
ISBN-13: 978-1-138-73743-3 (hbk)

Library of Congress Cataloging-in-Publication Data
Names: Karimi, Hamid Reza, editor.
Title: Offshore mechatronics systems engineering / editor, Hamid Reza Karimi.
Description: Boca Raton, FL : CRC Press, Taylor & Francis Group, [2018]	"A science publishers book."	Includes bibliographical references and index.
Identifiers: LCCN 2018015083	ISBN 9781138737433 (hardback : acid-free paper)
Subjects: LCSH: Offshore structures--Automatic control.	Mechatronics.	Remote submersibles.
Classification: LCC TC1665 .O294 2018	DDC 622/.33819--dc23
LC record available at https://lccn.loc.gov/2018015083

Visit the Taylor & Francis Web site at
http://www.taylorandfrancis.com

and the CRC Press Web site at
http://www.crcpress.com

To my wife
Baharak

Preface

With the rapid growth of offshore technology in various fields such as oil and gas industry, wind energy, robotics and logistics, large space structures and autonomous underwater vehicles, and some other advanced technologies, many researchers in academia and industry have focused on technology-based challenges raised in offshore environment. To this aim, this book is introducing certain novel theoretical or practical techniques for offshore mechatronics systems.

This book is targeting as a reference for graduate and postgraduate students and for researchers in all engineering disciplines, including mechanical engineering, electrical engineering and applied mathematics to explore the state-of-the-art techniques for solving problems of integrated modeling, control and supervision of complex offshore plants with collective safety and robustness. Thus, it shall be useful as a guidance for system engineering practitioners and system theoretic researchers alike, nowadays and in the future.

The main feature in common of all the encompassed contributions in this book is that of creating certain synergies of modeling, control, computing and mechanics in order to achieve not only robust plant system operation but also properties such as safety, cost and integrity while retaining desired performance quality. It is thus believed the present book provides an innovated insight into the state-of-the-art of applications aspects and theoretical understanding of complex offshore mechatronics systems that emerged recent years either via physical implementations or via extensive computer simulations in addition to sound innovated theoretical developments.

The book chapters are organized as separate contributions and listed according to the order of the list of contents as follows:

Maintenance Logistics of Offshore Wind Farms: Influence of Sea Waves on Maintenance Logistics Processes by *Thies Beinke, Jan-Hendrik Wesuls, Jan Rosenkranz, Moritz Quandt, Abderrahim Ait Alla, Markus Schwarz, Daniel Syga, Thomas Rieger, Matthias Lange, Holger Korte* and *Michael Freitag* introduces the water-bound logistics maintenance processes of offshore wind farms as well as the necessary information for an overall planning and control of the processes under consideration.

Providing Standardized Processes for Information and Material Flows of Offshore Wind Energy Logistics by *Thies Beinke, Moritz Quandt* and *Michael Freitag* proposes a process concept for the offshore wind energy industry that results from a detailed process mapping of all supply chain partners.

Estimation-Based Ocean Flow Field Reconstruction Using Profiling Floats by *Huazhen Fang, Raymond A. de Callafon* and *Jorge Cortés* studies the critical problem of ocean flow field estimation based on inexpensive profiling floats.

Modern and Traditional Applications of Rheological Fluids in Mechatronic and Robotic Systems by *Sylvester Sedem Djokoto, Dragašius Egidijus, Vytautas Jurenas,Ramutis Bansevicius* and *Shanker Ganesh Krishnamoorthy* presents modern and traditional applications of rheological fluids in mechatronics and robotic systems.

Parameter Identification and Damage Detection of Offshore Structural Systems by *Jian Zhang, Xiaomei Wang* and *Chan Ghee Koh* introduces the procedures of dynamic analysis for complete structure and substructure in both time domain and frequency domain with application to a jack-up platform as an example.

Wave Energy Converter Arrays for Electricity Generation with Time Domain Analysis by *Fuat Kara* studies both numerical and analytical analysis for wave interaction in an array system using two and four truncated vertical cylinder arrays.

Variable Structure Control via Coupled Surfaces for Control Effort Reduction in Remotely Operated Vehicles by *A. Baldini, L. Ciabattoni, A.A. Dyda, R. Felicetti, F. Ferracuti, A. Freddi, A. Monteri`u* and *D. Oskin* develops a coupled surfaces Variable Structure Control to solve the tracking problem for a remotely operated vehicle.

Marine Brushless A.C. Generator by *Zenghua Sun, Guichen Zhang* studies the principles and construction of marine brushless A.C. generator for synchronous machines.

Marine Main Switchboard by *Guichen Zhang, Zenghua Sun* introduces new concepts and technologies, identify potential impacts, and explore new design methods to improve the Low Voltage Main Switchboard and the High Voltage Main Switchboard.

Generator Automatic Control System by *Guichen Zhang, Zenghua Sun* introduces new concepts and technologies, identify potential impacts, and optimum load sharing control method to realize energy conservation, emission reduction and cost control.

Finally, I would like to express appreciation to all contributors for their excellent contributions to this book.

Milan   **Hamid Reza Karimi**
20 April, 2018

Contents

1

Maintenance Logistics of Offshore Wind Farms

Influence of Sea Waves on Maintenance Logistics Processes

Thies Beinke,[1,*] *Jan-Hendrik Wesuls,*[2] *Jan Rosenkranz,*[3]
Moritz Quandt,[1] *Abderrahim Ait Alla,*[1] *Markus Schwarz,*[4]
Daniel Syga,[3] *Thomas Rieger,*[4] *Matthias Lange,*[3] *Holger Korte*[2] and
Michael Freitag[1,5]

1. Introduction

The expansion of offshore wind power in Germany is steadily progressing. With the addition of 108 offshore wind turbines (OWTs) in the first half of 2017, a total output of approximately 5 GW was achieved. This corresponds to a total of 1,055

[1] BIBA - Bremer Institut für Produktion und Logistik GmbH at the University of Bremen, Bremen, Germany.
[2] Jade University, Department of Maritime and Logistics Studies, Elsfleth, Germany.
[3] Energy & meteo systems GmbH, Oldenburg, Germany.
[4] Cluetec GmbH, Karlsruhe, Germany.
[5] Faculty of Production Engineering, University of Bremen, Bremen, Germany.
* Corresponding author: ben@biba.uni-bremen.de

OWTs connected to the power grid (Lüers et al. 2017). The increasing use of offshore wind energy (OWE) leads to the need to maintain the increasing number of OWTs during their operating phase. The most important aim of these maintenance efforts is a high availability of OWT. This is impeded by limited reachability and high logistics costs due to the rough conditions at sea (Oelker et al. 2016, Burkhardt 2013). To solve this dilemma, maintenance strategies have to be optimized (German Federal Ministry Economic Affairs and Energy 2015). Particular challenges in the maintenance of OWTs in comparison to the maintenance of onshore wind power plants result from the weather-limited accessibility as well as from the coastal remote location of offshore wind farms (OWFs) with the related costs for personnel or material transport via special vessels or helicopters (Oelker et al. 2016).

The prevailing weather conditions on the transport route as well as at the site of operation directly affect the processes of service logistics. Weather-related limit values of Crew Transports Vessel (CTVs) and the safety and workability of service technicians must be guaranteed. Therefore, consideration of weather conditions, in particular, wind speed, wave height and wave direction play a decisive role in planning and execution of maintenance of OWE.

In the case of operators involved in the operational implementation of maintenance activities by OWT, such as the operating company and the service providers for the transport of service technicians and seaport operators, decentralized information management has hitherto prevailed. As a result, for both the transport route and the operational area, especially with regard to border weather conditions, all the relevant information to deployment information is not available, e.g., weather and sea predictions or ship availability.

This contribution examines the potential of increasing the transparency of information on the implementation of offshore service approaches. The following Section 2 presents a background for the logistics of maintenance of OWE. Therefore, it describes the process of planning, preparing, and carrying out service operations for the maintenance of OWE. It also enumerates the necessary information flows between project partners, the possibilities of information generation in the process, and economic relevance of a high level of information quality for operational service maintenance logistics of OWE. Section 3 presents how the CTV movement is dependent on the sea and waves. In addition to the analysis of the ship's movement, an approach to improve sea and wave forecasts is given in Section 4. A conclusion and an outlook on further research conclude this contribution (Section 5).

2. Maintenance Logistics of Offshore Wind Energy

With a duration of 20–25 years, the operating phase of an OWT is the longest phase in its life cycle. The operating costs during this phase amount to approximately 20% of the overall cost of an OWT. Therefore, the costs for maintenance constitute approximately 72% of these operating costs (Svoboda 2013). Due to this large share of operating costs, the current challenges of the industry are cost reduction and increase of turbine availability in order to secure efficiency and competitiveness (German Federal Ministry Economic Affairs and Energy 2015, Beinke and Quandt 2013, García-Márquez et al. 2012).

The maintenance work requires transportation of both materials and personnel (Oelker et al. 2013). The selection of the means of transport depends on the required transport capacities as well as on the actual weather conditions (Seiter et al. 2015, Franken et al. 2010). The weather conditions directly affect the usability such that the maintenance work can only be performed within specific limits which further depend on the respective means of transport (Quaschning 2013, Holbach and Stanik 2012).

2.1 Process of Offshore Service Logistics

The operative maintenance logistics process of OWE consists of three main processes; "planning of service application", "preparations for transfer" and "execution of transfer and maintenance". After the assignment of a service the offshore maintenance service provider (OMSP) by the OWF operator, the parties start planning this operation. In addition to the weather forecast, the maximum wave height is a major issue in the deployment decision, including information on personnel and resources. In case of unfavorable weather conditions, the execution of the service operation is postponed and the weather forecast is checked periodically.

The planning phase is followed by the preparations for transfer. For this purpose, the OMSP carries out an order-dependent load planning for each CTV which includes the type, time, and dimensions of the material and the scheduling of suitable loading aids.

Transfer and execution of the maintenance operations begin with an updated weather forecast to check the weather conditions immediately before the CTV leaves the seaport. If the weather conditions are beyond the defined limit values, the captain terminates the service operation after consultation with the OWF operator. Otherwise, the CTV is crewed with the service technicians. The transfer to the assigned OWT includes the wave height-dependent transfer of the service technicians from the CTV to the OWT. During the entire period of the execution of service tasks, the CTV stays in the OWF. In addition, the weather is continuously monitored. Changing weather conditions that predict wave height apart from the defined limit lead to the termination of the service operation. After completion or termination, the CTV picks up the service technicians and sails back to the base port.

It can be stated that a large number of influencing factors affect the operational logistics processes. In particular, the weather and wave conditions are a decisive criteria for the feasibility of service operations. Sea waves are the main reason for the

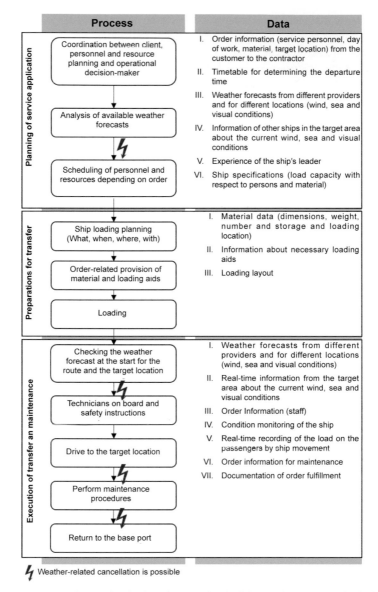

Figure 1. Process, data, and technology for operational offshore maintenance service logistics.

inaccessibility of OWT and are an influencing factor for the transfer. Therefore, sea waves in the operation area as well as on the transport routes need to be monitored. Rough sea conditions significantly affect the working of service technicians. The longer the prediction period, the more difficult and inaccurate is the estimation of the weather and sea waves. Figure 1 summarizes the described process of offshore maintenance service logistics as well as the necessary information flows for each part of the process.

2.2 Impact of Planning Uncertainties on Sea Conditions

Based on the processes and information requirements presented above, the question of the impact of insufficient information arises. In this context, Beinke et al. (2016) published a simulation study that examines the impact of the quality of sea state forecasts on the operative maintenance logistics of OWTs as well as the resulting limitations for power generation. The study focuses on the coordination process between the actors during the operational planning and considers different prediction qualities of wave height forecasts for the target area.

For operations conducted with service vessels, the operation decision is made exclusively based on the current weather forecast for the maximum wave height. Both players involved, the operation's company and the captain of the service vessel, have to decide whether the service operation should be conducted or not. In this simulation, we investigate two scenarios which allow the evaluation of the impact of wave forecast quality on the operative logistics process for OWT maintenance. To this end, the simulation considered an OWF with 60 OWT and an operating phase of 20 years. Due to different points in time at which the operating company and the captain make and verify their decision, this simulation model assumes different deviations from the actually measured wave height values for different weather forecast. It can be stated that the forecasts are more accurate if they are retrieved on the day of the operation rather than the day before.

The evaluation of the simulation results is based on three indicators: the availability of the wind turbines/the wind farm, the mean time to failure (MTTF) as well as the mean time to repair (MTTR). MTTF and MTTR are important parameters for quantifying system availability and time intervals between the occurrence and the repair of a system failure. In this context, the accessibility of the turbines or the weather restrictions plays an important role in reducing the MTTR. For the operative process, the number of trips as well as the number of wrong decisions due to the current wave height forecast is crucial.

$$MTTR = \sum_e^E \sum_w \frac{1}{|E|}(Tr_{ew} - T_{ew}) \tag{1}$$

$$MTTF = \sum_e^E \sum_w \frac{1}{|E|-1}(T_{(e+1)w} - Tr_{ew}) \tag{2}$$

Availability of the wind farm

$$Avail = \frac{MTTF}{MTTF + MTTR + D_A}$$ (3)

Therefore:

D_A	Duration of annual Inspection
E	Set of failure types
e	Index of a failure
SB	Time of shift start
SE	Time of the end layer
T_{ew}	Occurrence time of a failure of type e in the wind turbine w
Tr_{ew}	Repair time of failure e in the wind turbine w, where $Tr_{ew} \in [SB, SE]$

Subsequently, the results of this simulation study are presented in Table 1. For the presentation and discussion of the results, the table includes the scenario definition, number of trips per scenario, wrong decisions made based on the forecasts as well as the MTTR, MTTF, and the availability of the turbine.

The higher forecast accuracy in scenario β leads to a higher number of orders issued by the operating company on the day before the planned service operation. Therefore, the utilisation rate of suitable weather conditions reaches a significantly higher level. The reduction of non-initiated operation orders by approximately 78.5% in scenario β compared to scenario α clearly shows the impact of the forecast quality on the decision-making process. As a consequence, downtimes of wind turbines can be reduced which has a direct effect on the failure costs. For a 5 Megawatt OWT, the failure is equivalent to EUR 13,000 per day (Heidmann 2015).

Depending on the number of OWT per OWF and the prevailing weather conditions, high failure costs can occur in short time periods. In this case, we examined a wind farm size of 60 OWTs with an average OWT operating phase of

Table 1. Simulation results.

	Scenario α	Scenario β
Deviation of the forecast by the actual wave height on the previous day	± 20% ≙ ± 36 cm	± 10% ≙ ± 18 cm
Deviation of the forecast by the actual wave height on the working day	± 10% ≙ ± 18 cm	± 5 ≙ ± 9 cm
Number of trips	448.09	407.29
Wrong decisions based on the forecast	88.25	18.26
MTTR	142.11	163.79
MTTF	1740.64	1759.95
Availability in %	86.4669	85.6877

20 years. An increase in the wind farm availability of 0.79% leads to a reduction in the failure costs of approximately € 44.95 million over the wind farm operating phase. Furthermore, the lower MTTR in scenario β reduces the downtime per wind energy turbine by approximately 5.4 days per year.

Based on the economic significance of the quality of information for decision-making in operational offshore wind energy maintenance logistics, this contribution presents two approaches to raise information quality. This comprises the prediction of motions of CTVs and an approach to improve sea wave forecasts for different target destinations.

3. Wave Behaviour of the CTV

To ensure proper maintenance cycles and the safety of offshore personnel, real-time motion predictions for observer technologies within a logistic frame are needed. This requires a detailed knowledge of loads and motions of the transport ship for the proper dimensioning of efficient and safe systems. In addition to the accurate calculation of the structural movements, the calculation speed of a simulation is an important quality feature of this application; e.g., with regard to real-time constraints. To determine the kinematics model of a free-floating structure in a seaway, different approaches are commonly used. The difference between dry mechanics and maritime mechanics is based on additional hydrodynamic effects, namely the hydrodynamic added mass; e.g., Fossen (1994), Korte and Takagi (2004), and Kirchhoff (1869). The hydrodynamic mass force is an inertia force. The relative acceleration between incompressible fluid and structure induces a pressure field which results in a hydrodynamic force that is formulated as the product of the relative acceleration and a "virtual" mass. The size of the hydrodynamic added mass depends primarily on the direction of the structures movement and its geometry.

Due to the phenomenon of added or virtual masses, none-scalar and directed inertia values are necessary for describing translational and rotational motions. Probably, this results in an ordinary practice of describing floating structures within a body-fixed view, e.g., Kirchhoff motion equations of floating bodies (Kirchhoff 1869). A non-linear simulation is used to parameterise a linear real-time capable model of the ships' motion behaviour.

3.1 6DOF Motions of a Free-Floating Offshore Structure

In the following section, the six degrees of freedom of a ship are introduced. Figure 2 shows a sketch of a ship hull with a body-fixed coordinate system. A body-fixed coordinate system is fixed to the ship and moves with it. Due to advantages in determining the moments of inertia, the origin of the system is preferably located in the ship's centre of gravity. The motions of a ship are described in the body-

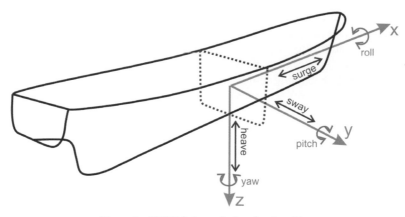

Figure 2. 6DOF Motions of a free-floating ship.

fixed coordinate system. However, the ship's position as well as its orientation is referenced in an earth-fixed coordinate system.

The translational motions are:

- Surge: Translation along the longitudinal x-axis
- Sway: Translation along the transversal y-axis
- Heave: Translation along the vertical z-axis

The rotational motions are:

- Roll: Rotation around the longitudinal x-axis
- Pitch: Rotation around the transversal y-axis
- Yaw: Rotation around the vertical z-axis

It should be noted that translational and rotational motions are not independent, they are rather coupled, i.e., a heave motion will induce a pitch motion and vice versa.

Alternatively it can be distinguished between horizontal (surge, sway, and yaw) and vertical (heave, roll, and pitch) motions, which results in a simplification of the motion equations as the coupling of forces and moments is neglected. The reason for this differentiation is the restoring force caused by gravitation and buoyancy. After a perturbation of the equilibrium, a ship always tends to return to it. This does not apply to horizontal motions. Here, a perturbation results in a drift of the free-floating structure away from its original position as well as a change of its heading. Typical perturbations are wind, waves and current. For a driven ship or a ship in dynamic-positioning mode, the ships actuators (propeller, rudder, thruster, etc.) control the horizontal motions.

3.2 Motion Equations in the Body-Fixed Reference System – Kirchhoff Equations

As already mentioned, hydrodynamic inertial effects in the form of hydrodynamic added mass have to be taken into account in marine applications. The mass matrix m in the traditionally used body-fixed reference system is defined in Equation 5. In the matrix, m is the physical mass and $m_{h,ij}$ is the direction-dependent content of the hydrodynamic inertia. The mass matrix is constant in the body-fixed frame. This applies analogously to the matrix of moments of inertia.

$$m_{i,j} = m + m_{h,ij} \tag{4}$$

$$\bar{m} = \begin{bmatrix} m + m_{h,11} & 0 & 0 \\ 0 & m + m_{h,22} & 0 \\ 0 & 0 & m + m_{h,33} \end{bmatrix} \tag{5}$$

In 1869, Kirchhoff published his work about the movement of a rotating body in a fluid (Kirchhoff 1869). In this work, he defined the motion equations of a floating body in a body-fixed reference system in analogy to Euler's gyro equations. The Kirchhoff equations are a system of three equations each of translation (Equation 6) and rotation (Equation 7).

$$\frac{d\vec{P}}{dt} + \vec{\omega} \times \vec{P} = (X,Y,Z)^{\mathrm{T}} \tag{6}$$

$$\frac{d\vec{L}}{dt} + \vec{\omega} \times \vec{L} + \vec{v} \times \vec{P} = (K,M,Z)^{\mathrm{T}} \tag{7}$$

The so-called living forces and moments depict the external forces and moments including hydrodynamic effect, weight, and buoyancy. Their effect is contained in the right-hand side of the above equations. The determination of these external forces and moments is especially difficult for the horizontal degrees of freedom with the lack of restoring forces. For the simulations in this work, a simplified model is used which considers weight, buoyancy, as well as potential damping. At present, viscous effects are neglected.

3.3 Simulation

3.3.1 Ship parameterisation

The simulated ship is a crew transfer vessel which is used for the transfer of offshore service staff in the German Bight. Figure 3 shows a CAD-model of the ship. It is a catamaran hull with length $L_{oA} = 22.0$ m, breadth $B = 8.3$ m, and mass

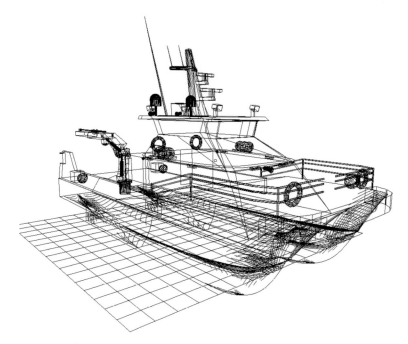

Figure 3. CAD-snapshot of the simulated Crew Transfer Vessel.

$m = 60\ t$. For the simulation, the hydrodynamic parameters such as hydrodynamic added masses, moments of inertia as well as the potential damping coefficients of the CTV are required. They were determined within the project "SOOP – Safe Offshore Operations" using the potential radiation and diffraction program WAMIT cf. Korte et al. (2012). The ship is discretised station-wise to calculate wave induced forces and moments. The calculated forces are buoyancy and weight as well as potential damping force.

3.3.2 Determining the Response Amplitude Operators (RAOs) of the ship

The motion analysis does not determine the motion functions analytically by means of a Laplace transformation since the motion behaviour of the vessel would have to be linearly represented. Rather, ship motions are systematically induced through specified seaways which were created by using analytical JONSWAP-spectra, cf. Hasselmann et al. (1973). Each simulation run covers a statistically relevant time period of 3 hours. To reduce computational load, the exciting seaways and response motions are linearized. During the simulation runs, the resulting ship motions are recorded and visually presented.

As the recorded ship motions are assumed to show a harmonic behaviour, they can be transferred from the time domain to the frequency domain by using the well-known Fourier transform. An input-output analysis of the sea-wave spectra and the response spectra leads to characteristic response amplitude operators (RAOs) of the ship, which represent a set of direction-dependent transfer functions of the ship motions. The simulation calculation creates data sets of the frequency dependent motion characteristic of the vessel for all the 6 degrees of freedom and for nineteen wave encounter angles ($\beta_0 = 0°–180°$) in total, where head seas are defined as an encounter angle of 180°. Due to the symmetry of the vessel, the remaining angles can be easily derived from these data sets. The ship's speed is set to 0 knots for all simulations. It has no peculiar velocity.

A tool, developed in MATLAB®/Simulink can simulate the motions of any kind of free-floating structure if geometry, mass, moments of inertia, and hydrodynamic properties of the structure are known. By adapting the wave spectra to real measurements, the ship motions can be determined directly from the simulation in any sea area, as long as the specific sea state data for this area is available.

3.3.3 Real-time transformation to an arbitrary seaway

The advantage of the chosen approach consists in describing the motion responses to exciting seaway through ship-specific transfer functions known as response amplitude operators (see above). In this way, the motion responses to any exciting seaway can be determined swiftly.

Figure 4 gives an operational sequence description to determine the heave motion of a ship in a given sea state by using response amplitude operators.

Taking into account the linear wave theory (Airy wave theory, cf. Airy 1841) the surface elevation in the time domain can be transferred into frequency domain (Figure 4, left column); i.e., the seaway is decomposed into a number of wave components using a Fourier transform algorithm. Each wave component is defined by its amplitude ζ_a, angular frequency $\omega = 2\pi/T$, and phase ε. All wave components form the Fourier spectrum $|F_\zeta(\omega)|$, the amplitude spectrum $\zeta_a(\omega)$, and the sea state spectrum $S(\omega)$, which can be converted into each other using the following relations, see Equations 8 and 9.

$$\zeta_\alpha(\omega) = \frac{|F_\zeta(\omega)| \cdot 2}{N}[m] \tag{8}$$

$$S(\omega) = \frac{\zeta_{a(\omega)}^2}{2\Delta\omega}[m^2 s] \tag{9}$$

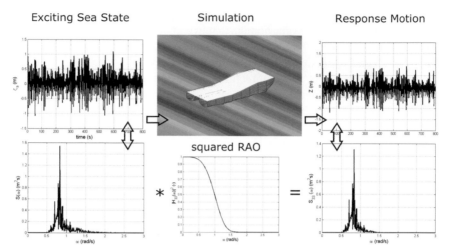

Figure 4. Operational sequence description of the motion analysis.

By multiplying the sea state spectrum $S(\omega)$ of the exciting seaway derived from weather forecasts with the squared RAO, which represents the frequency dependent motion characteristics of the vessel, each exciting wave component is superimposed by its related response component, see Equation 10; Figure 4, bottom row.

$$S_{i,i}(\omega) = H_i^2(\omega) \cdot S(\omega) \tag{10}$$

Analogous to the sea state spectrum, all frequency dependent system state responses from the response spectra $S_{i,\beta}(\omega)$ of motion for the six DOF and the examined wave encounter angle β_0. Subsequently, the frequency dependent response spectra can be transferred into the time domain by using an inverse Fourier transform algorithm, see Figure 4, right column.

It should be noted that derived seaways correlate with the statistical parameters of the sea state spectra from analyzed weather forecasts. However, this does not indicate the actual seaway during the observation period as phasing of the unique wave components is unknown. To acquire specific ship motions, one can modulate the phasing of the sea state spectrum until the desired sea states and responses are obtained. This operational sequence is a common procedure in maritime technologies; for a detailed description see Jacobsen (2005).

3.3.4 Motion results

The results of the motion analysis are shown in Figure 5 for one exemplary case. It shows the heave and pitch motions of the examined CTV in head seas (encounter

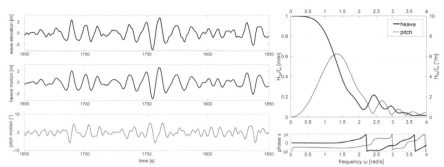

Figure 5. Calculated seaway and motions (left) and response amplitude operators (right).

angle $\beta_0 = 180°$). The seaway was generated by using the standard JONSWAP spectrum with a significant wave height $H_s = 3,0\ m$ and a peak period of $T_p = 12\ s$, which is superimposed with random phases. To generate a statistically relevant sea state, it was generated over a time period of 3 hours. The left hand side of Figure 5 shows a representative excerpt of this seaway in the time domain as well as the corresponding heave and pitch motions from which the maximum wave height and the maximum motions can be read directly. As expected, the heave motions follow the wave elevation, while the pitch motion follows the wave steepness.

The right-hand side of Figure 5 shows the absolute values as well as the phasing of the response amplitude operators for the heave and pitch motions used in this calculation. The phase shift between heave and pitch is easily recognizable.

In order to make the movement data more transparent to users and to reduce computational load, the data is displayed in a compact form. The simulation results are reduced to three degrees of freedom, heave, roll, and pitch since these are of the greatest relevance for the safe ship operation. Furthermore the consideration of possible encounter angles was adapted to users' requirements. Since deviations from the direct heading can occur in practice, the ships heading is extended by an aperture angle of $\pm 30°$ for the analysis to acquire a range of likely wave encounter angles, as is shown in Figure 6.

The expected maximum ship motions are hereafter calculated solely for the three degrees of freedom and for the range of wave encounter angles defined by the ships heading and the aperture angle.

4. Sea and Wave Forecast

Accurate weather predictions play a key role in the successful scheduling of offshore maintenance and service work. To be able to create reliable action planning, accurate predictions for upcoming meteorological situations such as thunderstorms, fog, and heavy gusts are required. Furthermore, reliable wave forecasts such as wave

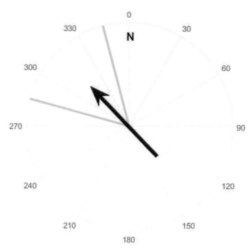

Figure 6. Ships heading and aperture angle.

height, direction, and frequency are crucial for the navigation to the wind turbines and especially for the process of the boat landing. This section focuses on the sea, and wave forecasts and explains them in detail.

After consultation with the management and ship's captain of an offshore transport service provider, it was decided to focus on a prediction horizon of 0–72 hours which corresponds to the average planning period. However, since the actual decisions about the feasibility of operations are made few hours before the scheduled start, a special emphasis is given to intraday predictions.

Seafaring conditions are a complex topic; waves alone can be described by a multitude of parameters. In discussion with an offshore transport service provider, we assessed that the significant wave height represents a solid approximation for the practicability of offshore operations. The significant wave height, commonly denoted as H_s describes the mean wave height of the highest one-third (33%) of the waves.

In general, numerical wind and wave forecasting models provide a lot of parameters describing the future state of the atmosphere and the ocean. The models are run several times a day. To handle the huge amount of data for the actual user, the data can be accessed through different interfaces. To facilitate an intuitive approach to the predictions, forecasts are provided as a customizable report, displaying the most important parameters as maps, time series, or tables. On the basis of communicated warning thresholds, on the one hand, the feasibility of offshore operations can be indicated directly in the report. On the other hand, the whole data set, especially relevant for automated systems, can be accessed by all major interfaces such as web services, SFTP, or e-mail.

4.1 Wave Forecast Method

Several providers operate complex weather models specialized in wave predictions. As the first step to creating the best possible wave prediction, we assessed the quality of a number of relevant forecast providers focusing on their performance in the North Sea. Next, we adjusted the predicted wave results to buoy data. Finally, the optimized wave predictions of the individual models were combined into a single final enhanced wave prediction.

4.1.1 Improvement of individual wave predictions

Within the scope of the project, buoy data containing wave measurements were available. The data contained high-resolution wave measurements for several years. With the insight of the data set, we analyzed the individual wave predictions and developed a correction strategy. For every prediction operator, an individual correction was realised.

4.1.2 Combination of individual wave predictions

Prediction models from different forecast providers differ in terms of the numerical model used, the parameters of the physical phenomena, and the assimilation data. Due to the diversity in the forecast creation process, anti-correlated forecast errors can be expected to occur. By combining the individual predictions, the anti-correlated errors are canceled out, thereby, increasing the overall forecast quality. The success of prediction combinations has been shown in different publications and has proven in operation usage (see Figure 7).

Mathematically, the combining process is a linear superposition of the corresponding wave predictions p_i (Equation 11). Every prediction receives a linear factor (model weights) α_i representing the model's influence on the resulting combined wave prediction p_c. N describes the total number of models used.

$$p_c = \sum_{i=1}^{N} \alpha_i * p_i \tag{11}$$

A correct parameter of the model weight α_i is crucial for a successful combination. To determine the optimal model weights, the historical forecast performance of the individual predictions was analysed. However, to exploit the complete potential of the forecast combination process, interactions between different predictions have to be taken into account and incorporated into the final model weights.

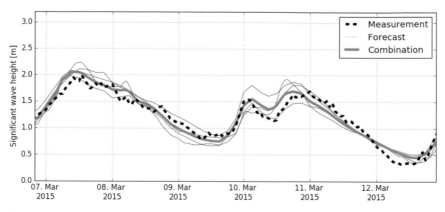

Figure 7. Wave forecasts of significant wave height for a location in the North Sea. Measurements obtained from buoy data.

4.1.3 Forecast quality

To evaluate the quality of the improved and combined wave predictions, a systematic analysis was conducted. This analysis focussed on the significant wave height, which is one of the most prominent parameters to describe seafaring conditions. To align the evaluation to the needs of offshore maintenance scheduling, the forecast data was divided into three predictions horizons: intraday, 1-day ahead and 2-day ahead. The intraday forecast horizon describes all forecasts with a horizon of 0–24 hours, the 1- and 2-day ahead predictions contain the time range of 24–48 and 48–72 hours respectively.

Overall, the optimized and combined wave prediction matches the buoy data remarkably well (see Figure 8). To describe the forecast quality statistically, the

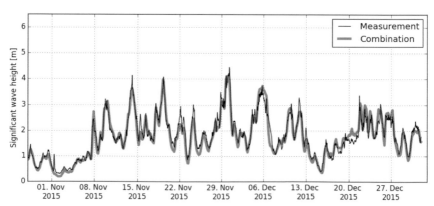

Figure 8. Wave prediction and buoy data for a location in the North Sea.

root-mean-square-error (RMSE), mean-absolute-error (MAE), and systematic error (BIAS) were determined for respective forecast horizons (see Table 2). On an average, intraday predictions show an error of 17.8 cm and a systematic error of 3 cm.

Table 2. Forecast quality of optimized and combined wave predictions. All units are displayed in meters.

	RMSE	MAE	BIAS
Intraday	0.264	0.178	0.030
1-day ahead	0.280	0.189	0.021
2-day ahead	0.333	0.217	0.043

4.2 Probability of Exceedance

Sea state is an extremely complex system and, as a direct consequence, difficult to predict. Therefore, all open sea wave prognosis is bound to contain at least minor inaccuracies. As a consequence, assessing and analyzing possible uncertainty is necessary for a responsible risk management.

To assess and handle risks originating from forecast inaccuracies, the wave parameter Probability of Exceedance (PoE) was developed and provided for all participants of the project. The parameter describes the probability of the occurrence of a wave height above a pre-defined threshold, thereby, presenting an intuitive approach to occurring forecast uncertainty. The calculation of the PoE is based on the analysis of historical wave measurements and forecasts and can be determined for different wave parameters. Next, to the original wave forecast, the prediction horizon is of great importance for the calculation of the PoE value to reflect the increase in uncertainties with increasing forecast horizon.

5. Conclusion and Outlook

In this contribution, we have introduced the water-bound logistics maintenance processes of OWFs. The processes of planning of service application, preparations for transfer, and execution of transfer and maintenance of the maintenance logistics with CTVs followed this first introduction. It included the explanation required for the overall planning and control of the mentioned processes. Based on this process description, it is clear that sea and waves have a significant influence on the processes. The economic impact of the prediction of sea and waves motivated the following presentation of approaches that determine the CTVs motion and improves the predictions of sea and waves forecasts.

A method to predict the characteristic motions of a CTV in a nearly real-time environment and to couple these predictions with weather forecasts has been shown. Although these predictions rely heavily on a stochastic approach and cannot predict single waves and therefore cannot predict the exact motions, the overall expected sea state and motion behaviour has been recorded.

High-quality sea and wave forecasts are essential for further processes such as the planning and the execution of the CTV transfer. Especially the calculation of the predicted CTV motion is highly dependent on the accuracy of the wave forecasts. Due to error propagation, even the best "wave spectrum to CTV motion transfer" process is useless without high-quality wave spectrum forecasts as its most important input parameter.

Wave forecast quality can be improved significantly by combining the output of several independent numerical weather predictions. To verify these prediction models, it is necessary to compare it to buoy measurement data. At the moment, the amount of different buoy measurements in the North Sea is quite low. In future, a higher number of buoys with a wave measurement will lead to more precise wave forecasts.

It is clear that in addition to an optimal forecast of sea and waves and the knowledge of the movement of CTVs, an inclusion of all the information in an integrated planning and control tool is needed. Further research activities of the authors include the development of the aforementioned approaches and the improvement of planning and control of maintenance logistics.

Acknowledgements

The authors would like to thank the German Federal Ministry of Economic Affairs and Energy for their support to the project IeK - Information System for near real-time Logistics (funding code 16KN021723ZIM-KF).

References

Airy, G.B. (1841). Tides and waves. In Hugh James Rose, et al. Encyclopædia Metropolitana. Mixed Sciences. (published 1817–1845). Also: Trigonometry, On the Figure of the Earth, Tides and Waves, 396 pp.

Beinke, T. and Quandt, M. (2013). Globalisierung des Windenergiemarkts - Potenziale für Praxis und Forschung – Analyse anhand zweier exemplarischer Beispiele. *In*: Industrie Management, 29(2013)5: 49–52.

Beinke, T., Quandt, M., Ait Alla, A., Freitag, M., Rieger. T. (2017). Information system for the coordination of offshore wind energy maintenance operations under consideration of dynamic influences. *In*: International Journal of e-Navigation and Maritime Economy, 8/2017, pp. 48-59.

Burkhardt, C. (2013). Logistik- und Wartungskonzepte. *In*: Böttcher J. (Hrsg.): Handbuch Offshore-Windenergie. Rechtliche, technische und wirtschaftlicheAspekte. Oldenbourg, München.

Fossen, T.I. (1994). Handbook of Marine Craft Hydrodynamics and Motion Control. 1st ed. (UK): John Wiley & Sons Chichester.

Franken, M., Trechow, P., Kaluza, M. and Matook, S. (2010). Offshore. Service & Maintenance.

GarcíaMárquez, F.P., Tobias, A.M., Pinar Pérez, J.M. and Papaelias, M. (2012). Condition monitoring of wind turbines. Techniques and methods. Renewable Energy, 46: 169–178.

German Federal Ministry Economic Affairs and Energy (2015). Offshore-Windenergie. Die Energiewende - ein gutes Stück Arbeit. Ein Überblick über die Aktivitäten in Deutschland. URL: http://www.bmwi.de/BMWi/Redaktion/PDF/Publikationen/offshore-windenergie,property=pd f,bereich=bmwi2012,sprache=de,rwb=true.pdf, zuletzt geprüft am 28.07.2017.

Hasselmann, K., Barnett, T.P., Bouws, E., Carlson, H., Cartwright, D.E., Enke, K., Ewing, J.A., Gienapp, H., Hasselmann, D.E., Kruseman, P., Meerburg, A., Mller, P., Olbers, D.J., Richter, K., Sell, W. and Walden, H. (1973). Measurements of wind-wave growth and swell decay during the Joint North Sea Wave Project (JONSWAP). In Ergänzungsheft zur Deutschen Hydrographischen Zeitschrift. Reihe, A(8) (Nr. 12), p. 95.

Heidmann, R. (2015). Windenergie und Logistik. Losgröße 1: Logistikmanagement im Ma-schinen- und Anlagenbau mit geringen Losgrößen. Beuth Verlag GmbH, Berlin, Wien, Zürich.

Holbach, G. and Stanik, C. (2012). Betrieb von Offshore-Windparks. Maritime Dienstleistungspotenziale. Internationales Verkehrswesen 64(6): S. 31–33.

Jacobsen, K. (2005). Hydrodynamisch gekoppelte Mehrkörpersystemeim Seegang Bewegungs-simulationen im Frequenz-und Zeitbereich, Dissertation, TU Berlin.

Kirchhoff, G. (1869). Über die Bewegung eines Rotationskörpers in einer Flüssigkeit. Crelles Journal, 71: 237–273.

Korte, H. and Takagi, T. (2004). Dynamic motion calculation of a flexible structure using inertia transformation algorithm. pp. 1–11. In: Workshop on Fishing and Marine Production Technology; 9.-10. August; Trondheim. NTNU.

Korte, H., Ihmels, I., Richter, J., Zerhusen, B. and Hahn, A. (2012). Offshore training simulations. pp. 37–42. In: 9th IFAC Manoeuvring and Control of Marine Crafts; 19.-21. September 2012; Arenzano (11). IFAC. DOI: ISSN 1474-6670.

Lüers, Silke; Wallasch, Anna-Kathrin; Vorgelsang, Kerstin: Status des Offshore-Windenergieausbaus in Deutschland. 1. Halbjahr 2017. Hg. v. Deutsche WindGuard. URL: http://www.windguard. de/_Resources/Persistent/58f54330828c2c6dff921aa9e457fd27fca0eb5a/Factsheet-Status-Offshore-Windenergieausbau-Jahr-1.-Hj.-2017.pdf, zuletzt geprüft am 28.07.2017.

Oelker, S., Ait Alla, A., Lewandowski, M. and Freitag, M. (2016). Planning of maintenance resources for the service of offshore wind turbines by means of simulation. pp. 303–312. In: Proceedings of the 5th International Conference on Dynamics in Logistics (LDIC 2016), Springer.

Quaschning, V. (2013). Erneuerbare Energien und Klimaschutz. Hintergründe - Techniken und Planung - Ökonomie und Ökologie - Energiewende. Carl Hanser Verlag; Hanser, München.

Seiter, M., Rusch, M. and Stanik, C. (2015). Maritime Dienstleistungen. Potenziale und Herausforderungen im Betrieb von Offshore-Windparks. Springer Gabler, Wiesbaden.

Svoboda, P. (2013). Betriebskosten als Werttreiber von Windenergieanlagen – aktueller Stand und Entwicklungen. Energiewirtschaftliche Tagesfragen, 63(5): 34–38.

19

2

Providing Standardized Processes for Information and Material Flows of Offshore Wind Energy Logistics

Thies Beinke,[1,*] *Moritz Quandt*[1] and *Michael Freitag*[1,2]

1. Introduction—Construction Phase Logistics for Offshore Wind Energy

Offshore wind energy represents a key technology within the energy revolution (Federal Ministry for Economic Affairs and Energy 2015, Hau 2014). This is due to the high supply potential of wind out at sea, the resulting high number of full-load hours, and the favorable power plant characteristics that offshore wind energy offers (Hau 2014, Rohrig et al. 2013). In view of Germany's decision to phase out nuclear energy, offshore wind energy represents a form of power generation that is capable of supplying base load, making it a key component in the future energy mix (German Federal Government 2010). The need to optimize and reduce costs in all areas of the supply chain of offshore wind energy is reasoned by the following aspects: the specific challenges of offshore wind energy as well as the competitive

[1] BIBA - Bremer Institut für Produktion und Logistik GmbH at the University of Bremen, Bremen, Germany.
[2] Faculty of Production Engineering, University of Bremen, Bremen, Germany.
* Corresponding author: ben@biba.uni-bremen.de

situation with conventional and other renewable energy sources (Federal Ministry for Economic Affairs and Energy 2015). Given that the cost-saving potential for production and installation logistics was estimated at 5.7–9% and 3.6–5% respectively back in 2013, there is a need for optimisation in these areas (Briese 2016, Hobohm et al. 2013).

Logistics is defined as the market-oriented, integrated planning, design, processing, and control of the entire material flow and associated information flow between a company and its suppliers, within a company as well as between a company and its customers (Schulte 2013). Maritime logistics covers the onshore and offshore transport chain, its fields of design with respect to shipping and maritime transport, land and maritime hinterland transport, and the ports including the ways they create value and their necessary infrastructure and suprastructure (Klaus et al. 2012). In the context of offshore wind energy, Thoben et al. (2014) describe logistics as a collection of different concepts, processes, and technologies used for material and IT-related coordination of agents throughout the entire life cycle. They put forward logistics as a critical factor for success. Within the construction and operation phases of the supply chain, logistics assumes a key role, making up around 25% of the total investment costs of an offshore wind farm (Hau 2014, Heidmann 2010). Construction logistics accounts for the lion's share of the costs (Reichert et al. 2012).

The range of topics relating to construction logistics is wide and varied. In addition to site logistics, transport, storage, and handling, the field comprises other areas such as engineering, cost, and transfer of risk arrangements. Additionally, it also includes the context of a country's exclusive economic zone and strategic development as a way of increasing resource efficiency (Beinke et al. 2015). This serves to illustrate that storage, transport, and handling processes are only a small part of offshore wind energy logistics when it comes to process design and control of material and information flows (Heidmann 2015).

1.1 Challenges and Requirements

The challenges of construction and construction logistics are primarily described in the literature with respect to the specified environmental conditions and technological challenges. The increasing distance from the coast, the associated water depth, and compliance with nature conservation legislation; along with the challenge to use suitable weather conditions represent the environmental factors (Hau 2014, Briese and Westhäuser 2013, Gasch and Twele 2013). The wind constitutes the greatest challenge as this is a limiting factor for the conduction of a variety of processes on land and especially at sea (Stohlmeyer and Ondraczek 2013).

Based on these challenges, requirements of and for offshore wind energy logistics are formulated as follows: on this matter, Burckhardt (2013) describes how new approaches to logistics should be developed as a result of natural circumstances. Briese and Westhäuser (2013) explain further that the requirements can be met primarily by standardizing processes, developing them to achieve series production, and ultimately by professionalizing the offshore wind energy industry. Regarding the logistics processes, this leads to developing sophisticated and certified logistics concepts and establishing corresponding logistics processes (Briese and Westhäuser 2013). In order to reduce the high costs of construction and construction logistics, the need for close cooperation between installers and ports as well as improvements in weather forecasting to prevent unsafe installation processes are also being addressed (Gille 2011).

According to Stohlmeyer and Ondraczek (2013), the weather represents the greatest risk to the completion of a project. Many essential offshore tasks may only be carried out up to a defined weather limit in order to ensure the safety of people and materials. Given that the weather is a disturbance variable that cannot be scheduled or influenced, it is clear that weather conditions and the waves that they cause have a considerable impact on the planned completion of a project. Both the wind and the waves define the available weather windows that are necessary for transport and installation (Ait Alla et al. 2016, Stohlmeyer and Ondraczek 2013). In the context of construction logistics, the construction concept, technology used, safety at sea, and the experience of the parties involved also represent risks in terms of whether a wind farm will be completed or not. The consequences of any of these completion risks can be a loss of earnings or sales, liquidity problems, and overspending on investment (Stohlmeyer and Ondraczek 2013).

Based on these findings, a detailed view of the construction phase of offshore wind energy turbines suggests itself. Thus, there can be substantial cost saving of the construction phase by standardizing work processes. The authors describe the supply chain network of offshore wind energy construction in the following section to provide a basis for the detailed process mapping in Chapter 3.

1.2 The Construction Phase Supply Chain for Offshore Wind Energy

The supply chain for production and construction is characterised by a directed material flow from the source (manufacturer) to the sink (installation site at sea) (Görges et al. 2014). The supply chain for the construction phase is characterised by natural conditions such as the port infrastructure or geographical position of the wind farm, the main components of a wind power plant (nacelle, tower, foundation structure, and rotor blades), as well as the types and number of vessels used (Ait Alla et al. 2016, Moccia et al. 2011).

Ait Alla et al. (2016) outline the supply chain for the construction phase of offshore wind energy as follows: the component manufacturers of the four main components of an offshore wind power plant form the starting point of the supply chain. The finished components are transported to the port by the logistics service provider. The plant components are held at the port, collected by the installer, transported to the wind farm and then installed. Here, the installer acts as the customer at the end of the supply chain. This supply chain system for a wind farm is summarized in Figure 1.

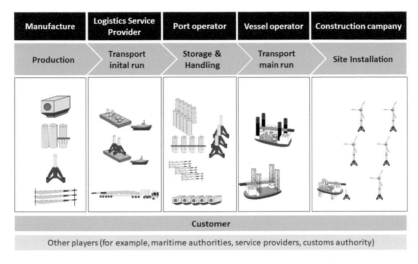

Figure 1. Logistics network of an offshore wind farm project (Quandt et al. 2017).

Due to the inter-company coordination over the offshore wind energy supply chain, the information quality plays a major role in efficient workflows. As already mentioned, the weather conditions are of vital importance for the construction processes. Therefore, the following chapter presents approaches for the inclusion of weather conditions in planning and control of offshore construction processes.

1.3 Planning and Control of Weather-Dependent Logistics Processes in the Context of Offshore Wind Construction

Based on conventional site production, frameworks are provided in the literature, which address the subject of planning and control. For example, Lee et al. (2006) consider strategic and operational aspects by means of scenario simulation using system dynamics, a multi-agent system, and the critical path method; and describe a highly dynamic planning approach. This is motivated by a variety of influencing factors and their dependencies. The consideration of weather-related disturbance variables in the planning of processes, as is essential in applications such as offshore wind energy, can be incorporated into different approaches for various industries. McCrea et al. (2008) describe this in a probabilistic framework with respect to weather-based re-routing and delays for aircraft. Further approaches which take weather conditions into account include planning models for scenario simulations and linear optimizations (for example, Ferrer et al. 2008, Le Gal et al. 2009).

The context of construction logistics for offshore wind energy and the influence of weather conditions is reviewed by, Tyapin et al. (2011), Scholz-Reiter et al. (2011) and Lütjen and Karimi (2012). Tyapin et al. (2011) present a study on the determination of suitable weather windows for the installation of an offshore wind power plant based on Markov theory and a Monte Carlo simulation. Scholz-Reiter et al. (2011) put forward a heuristic method for planning the installation process which also takes weather conditions and transport capacities into account. Lütjen and Karimi (2012) describe an approach that incorporates inventory control and supply of the port. In this context, Ait Alla et al. (2016) present a simulation study which considers the influence of information sharing in construction logistics for offshore wind energy. More works in the context of offshore wind energy can be found in relation to the operation of the plants. One can also refer to literature including Kovácsa et al. (2011), Zhang et al. (2012), and Oelker et al. (2016).

Based on the described challenges and requirements, the supply chain network of offshore wind energy construction as well as the preliminary scientific work, the need of a detailed consideration of the work processes of the entire logistics network becomes evident. Previous research has shown that the consideration of weather conditions for planning and control of construction processes provides useful information on the strategical, tactical, and operative level. On the contrary, preliminary approaches based on supply chain and weather models are on a high level of abstraction. Therefore, this contribution provides standardized logistics processes for the entire supply chain of offshore wind energy construction which can serve as a basis for further optimization of planning and control in this field of application.

2. Research Approach

Based on the requirements described above, the aim of this contribution is to develop a process concept that provides standardized material and information flow within the logistics network of the offshore wind energy industry. To develop this concept, the following approach was chosen (Figure 2).

The first step was to gain expert knowledge of the offshore wind energy industry. This was achieved by gathering detailed process information from all partners of a logistics network of a specific offshore wind farm project. The information about the current processes of the network partners was collected and verified by interviews and workshops with experts of the individual companies.

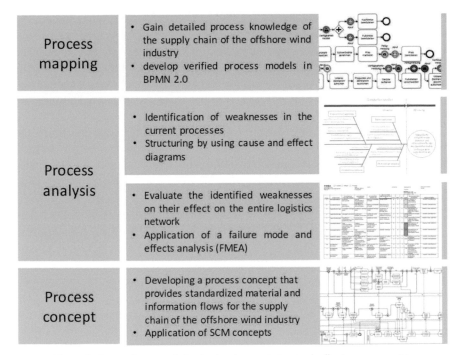

Process mapping	• Gain detailed process knowledge of the supply chain of the offshore wind industry • develop verified process models in BPMN 2.0	
Process analysis	• Identification of weaknesses in the current processes • Structuring by using cause and effect diagrams	
	• Evaluate the identified weaknesses on their effect on the entire logistics network • Application of a failure mode and effects analysis (FMEA)	
Process concept	• Developing a process concept that provides standardized material and information flows for the supply chain of the offshore wind industry • Application of SCM concepts	

Figure 2. Research approach for the development of a standardized process model.

At the end of the process mapping cycle, the current processes of the entire logistics network were documented in detailed BPMN (Business Process Model and Notation) 2.0 models.

The processes of the network partners were analyzed on weaknesses that were classified in consistent categories within the following step. This structured collection was achieved by using a cause-and-effect diagram for each network partner. A cause-and-effect diagram helps to actively seek causes for a defined effect and provides solution approaches for a specific problem (Ishikawa 1986).

Subsequently, the identified weaknesses of the individual network partners were evaluated on their relevance and influence on the entire supply chain of an offshore wind farm project. The evaluation of the weaknesses was conducted by applying a Failure Mode and Effect Analysis (FMEA) that provides the prioritization of potential failures and enables to identify system inherent vulnerabilities (Sankar and Prabhu 2001). It is helpful to use a FMEA during process optimization as information is systematically arranged within a given structure. Therefore, it can be used by teams to communicate about the weaknesses and to cooperatively develop countermeasures as well. As a result, the most relevant weaknesses for

25

the entire logistics network were identified and goals for the development of the process concept could be derived.

Based on the three described steps, the process concept was developed. The concept includes all relevant stakeholders of the logistics network. Besides the material flow, the information flows over the entire network were standardized. Furthermore, the process concept reveals options to automate selected information flows.

In the following sections, the process mapping, the process analysis, and the FMEA are explained using the example of the nacelle manufacturer. As already mentioned, these steps were conducted for the entire logistics network that includes the component manufacturers, the logistics service provider, a port operator, and the construction company.

3. Process Development for the Offshore Wind Energy Installation

3.1 Process Mapping

The process mapping was conducted for all partners of the supply chain of the offshore wind energy industry. For example, a selected process model is shown and explained. Due to the complexity of the actual production process of the

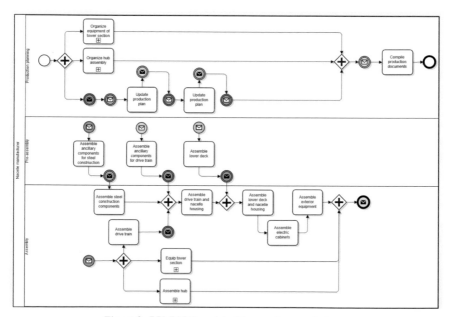

Figure 3. BPMN 2.0 model of the nacelle manufacturer.

selected nacelle manufacturer, the process model is shown in a simplified version that comprises the key processes (see Figure 3).

The process model of the nacelle manufacturer is divided into three lanes. Each lane represents one department that takes part in the production process. The production planning department is responsible for the coordination of the production process. The pre-assembly department produces parts for the assembly department that are required at defined production stages.

The production process is divided into three separate components. The production planning department simultaneously organizes the equipment of the bottommost tower section, the assembly of the hub, and the assembly of the nacelle. The illustrated process model (Figure 3) only shows the detailed process of the assembly of the nacelle. Both the equipment of the tower section and the assembly of the hub are shown as sub-processes that can be refined in a more detailed process model.

The assembly of the nacelle starts with status-signals from the production planning department to the pre-assembly as well as to the assembly department. These status-signals make sure that the required material is supplied and the production starts according to the production schedule. A status-signal is sent back to the production planning department as soon as a production stage is completed. The production plan is updated by the production planning department. Thus, the current status of the assembly departments can be gradually monitored.

In connection with the detailed process models of the other supply chain partners, the process mapping offers comprehensive process knowledge of an offshore wind farm project and, therefore, a viable basis for the identification of potential for improvement.

3.2 Process Analysis

The development of improved logistics processes is based on the results of a profound process analysis including all documented processes from logistics networks partners.

The following section shows the results of the respective process analysis of the production of the nacelle, hub, and the equipment of the lower tower section. The analysis uses a cause-and-effect diagram to visualize the main categories of weaknesses at each of the five branches of the diagram. The main branches divide into sub-branches that include different weaknesses of the specific category. Figure 4 shows the resulting cause-and-effect diagram from analyzing the processes of the nacelle manufacturer. The effect that results from the weaknesses causes an insufficient integration of material and information flows into the entire logistics network.

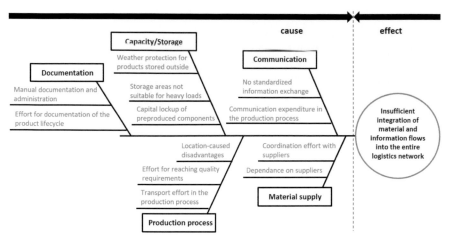

Figure 4. Cause-and-effect diagram of the nacelle manufacturer.

The main weaknesses occur within the dimensions of documentation, capacity/storage, communication, material supply, and within the production process. The processes of the nacelle manufacturer have to be documented to meet customers' requirements regarding quality standards, e.g., within the production. Most of the documents are generated and administrated manually. The complete product lifecycle of a component is documented which causes high efforts and costs. The integration of information technology to support the documentation would help to decrease connected efforts and standardize information flows.

The dimension capacity/storage show weaknesses that result from the weight and dimensions of the components. The components lock up capital as they partly need to be stored for months up to years. Additionally, the components need weather protection during storage if they are stored outside. As the components are extremely heavy, the subsurface of the storage space needs to be suitable for the heavy loadings.

Although communication is a fundamental process within every logistics and production network, weaknesses occur within this dimension. Regarding the producer of nacelle and hub, the analysis shows that the exchange of information is not standardized. Furthermore, communication causes high expenditures within the production process as a whole.

The supply of material causes coordination effort with the different suppliers. The dependence on these suppliers represents a central weakness for the producer.

The last dimension includes the production process. The main weaknesses are connected with the location of the production facilities of the nacelle manufacturer. The resulting transports cause high costs and efforts that manifest themselves in additional working hours and requisite tools and equipments. Moreover, the quality requirements for the production result in high additional effort to reach the requested durability of the components for the entire life cycle in the harsh offshore environment.

The second method used for the process analysis was Failure Mode and Effect Analysis (FMEA). This method was used to evaluate the weaknesses regarding their relevance for the entire supply chain network. Figure 5 shows an excerpt of the FMEA conducted for the process weaknesses of the nacelle manufacturer. The method was used to systematically analyze the previously identified weaknesses and to develop countermeasures for each one of it. The weaknesses in the logistics network were ranked by providing a risk priority number (RPN) for every potential failure. The RPN is a product of the probabilities of the occurrence, severity, and the likelihood of a specific weakness (Sankar and Prabhu 2001). Furthermore, the vulnerabilities were valued using a traffic light system depending on the RPN (green = low impact, yellow = moderate impact, red = high impact).

The most critical weaknesses identified by the different FMEA's were the insufficient communication between the network partners, inept infrastructure for the dimensions, weights of components of an offshore wind energy turbine, and weather influences on transport and construction processes.

Based on the detailed process analysis, objectives for the development of the process concept could be derived. The most important goal that had to be considered in the process concept was the provision of standardized information flows over the

Pos.	FUNCTION	POTENTIAL FAILURE MODE	POTENTIAL CAUSES	POTENTIAL EFFECTS	DETECTION METHOD/ACTION	OCC	SEV	DET	RPN	Recommended Action(s)	Responsibility
										chain	
4	capacity/storage	damaged products	insufficient weather protection	additional costs and work	improvement of weather protection	4	9	1	36	sheltered storage areas	nacelle manufacturer
5	capacity/storage	insufficient coordination of the production plan	uneconomic capital lockup	additional costs	modification of future contracts	8	2	8	128	optimisation of the material flows within the supply chain	supply chain partners
6	communication	high communication expenditure with supply chain partners	missing communication standard	suboptimal coordination of supply chain processes	regular meetings	9	7	8	504	development of standardized information flows	supply chain partners
7	communication	missing information in the production process	high communication expenditure in the production process	people-based knowledge	weekly meetings	9	6	8	432	development of standardized information flows	nacelle manufacturer
8	material supply	delay in delivery of essential parts	high dependency on suppliers	delay in production	search for further suppliers	5	8	3	120	Applying sourcing strategies	nacelle manufacturer
9	material supply	problems of suppliers to reach the requested quality	high communication expenditure with suppliers	additional work load	quality management for suppliers	4	8	5	160	Applying sourcing strategies	nacelle manufacturer
10	production	limited production	insufficient heavy-	enforcement of a	rent further	8	9	2	144	expansion of the	nacelle manufacturer

Figure 5. FMEA for the weaknesses of the nacelle manufacturer.

29

entire logistics network. Thus, an improved coordination between the supply chain partners, especially, on the numerous interfaces between network partners could be reached. A second important goal was to improve the coordination of material flows to reduce required storage areas throughout the supply chain. Furthermore, weather influences on the material flows and individual processes had to be included. This goal is directly connected with the need to improve the prediction quality of weather forecasts by documentation and analysis of operative data. Further goals were to reach high process stability in connection with the development of back-up processes due to weather conditions or occurring damages of equipment or components and the development of consistent quality standards for the entire logistics network with the participation of all supply chain partners.

3.3 Process Concept for Operational Planning and Control

The objective of the design of operational offshore wind energy logistics planning and control is to create transparency with regard to detailed project progress throughout the logistics network, establish support for the series production and construct offshore wind power plants, and to allocate resources efficiently. It is clear that these challenges can only be met by considering the flows of material and information across the network.

In accordance with the defined objectives, the process concept was developed. As the communication between the supply chain partners is of vital importance, an information interface has been implemented that acts as a central communication platform. The information interface offers the possibility to share information over the entire logistics network which improves process synchronization and, hence, fluctuations in output can be reduced. Furthermore, the improved transparency offered by shared information leads to a higher quality of the supply chain processes (Slack et al. 2010).

An overview of the involved supply chain partners and the information flows (shown as arrows) needed during the execution of an offshore wind farm project is illustrated in Figure 6. The involved partners are: the operator of the construction vessel; the different main component manufacturers that supply tower sections, nacelle, and hub; rotorblades; and the foundation structures. The logistics service provider is responsible for the transports from the main component manufacturers to the port of shipment and the handling and storage of the components in the port.

Based on retrograde scheduling, an initial schedule is developed collaboratively. The resulting schedule establishes limits for the completion of individual components. Once this initial schedule has been provided to all the parties involved in the network, this marks the end of this phase.

Notifying the time of completion of individual components enables the finished components to be incorporated into the overall transport plan promptly. Delays in

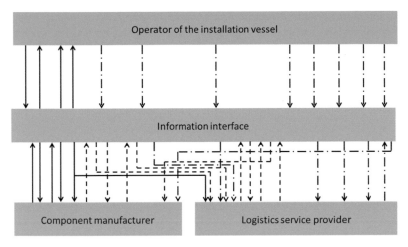

Figure 6. Outline process concept for the operational planning and control instrument (Bodenstab et al. 2014).

the production process result in the schedule being adapted; this is then provided to all partners involved. This enables any problem that occurs in the production processes to be responded to at an early stage. Before beginning the construction of individual components, the weather window for construction is checked. By providing information about changes, it is possible to adapt any deadlines to the new situation. Based on the status information during the construction of the wind farm, the schedule of supply to the port is constantly updated and made available to everyone involved via web-based planning and control system.

The third phase of information flow relates to status information during the construction of the wind farm. By constantly updating the announced arrival times of the installation vessels in the port, it is possible to forecast the demand for the components based on the schedule. Problems in the construction sequence can require logistics and production processes throughout the entire network to be rescheduled and controlled. Once the weather data has been called up, the logistics service provider specifies a collection date and uses the web-based system to intimate this to the component manufacturer. Further information that is relevant to planning and control in the supply chain is generated when the construction phase is complete, i.e., once the logistics service provider has finished shipping, upon leaving the construction field, when entering the port, and after mooring

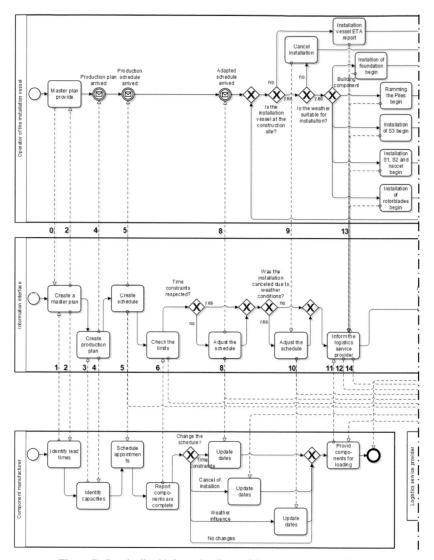

Figure 7. Standardized information flows of the process concept - Part 1.

the installation vessel. A detailed description of the individual information flows follows in Figures 7 and 8.

Figure 7 illustrates the first part of the standardized information flow between the web-based tool in the center of the graph and the main component manufacturers

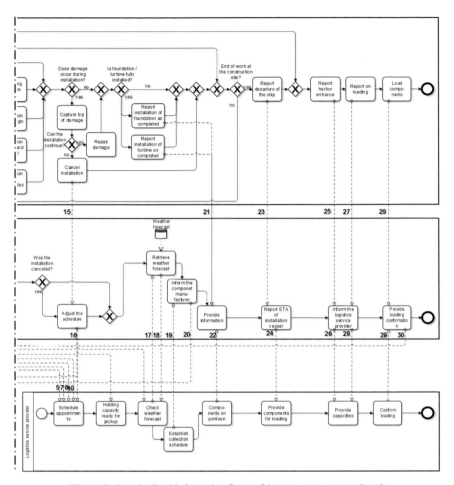

Figure 8. Standardized information flows of the process concept - Part 2.

on the bottom left. The operator of the construction vessel is illustrated in a collapsed pool at the top and the logistics service provider is shown at the bottom right.

The process starts with backward scheduling of the production planning of the individual main component manufacturers. The construction vessel, which is the most expensive link in the supply chain of the offshore wind industry, and the limited storage area for the components are the main reasons for using backward scheduling. Therefore, the supply of the components has to be adjusted to the schedule of the construction vessel. By using backward scheduling, the planning is conducted based on the due date of the last operation. Each individual process is planned backward until the first operation is scheduled. The main advantage of this approach is that the inventory of finished goods is minimized. However, it offers little margins for unplanned delays (Swamidass 2000, Gudehus and Kotzab 2012). That leads to the use of backward scheduling; in this individual case to set due dates for the production of components but it leaves the production planning to each individual component manufacturer.

The operator of the construction vessel that represents the construction company of an offshore wind farm provides the initial time schedule for the construction of the processed project (0). Additionally, the main component manufacturers provide the average lead times for the production of their components (1). On this basis, a first production schedule is generated and published (2) that is extended to a production plan as soon as the component manufacturers have submitted their production capacities (3) for the project duration. After submitting this production plan (4), an adjusted production schedule is generated and made available for all supply chain partners (5) that enhances the production plan by synchronizing the production processes of all component manufacturers. This production schedule sets specific limit values for the production of individual components that allow each component manufacturer the individual planning of their production but significantly contribute to a higher transparency within the entire logistics network. The next information flow is generated by the component manufacturers when a component leaves the production and a notice of completion is reported to the web-based system (6). The notices of completion are forwarded to the logistics service provider (7) to guarantee a prompt execution of the transport planning by the logistics service provider. If a disturbance occurs in the production process, the production plan is adjusted according to the reported delays to all supply chain partners (8). The reported status information enables all partners to react immediately on occurring delays in the production process.

Prior to the beginning of the installation of individual components, the weather conditions for the specific installation stage are checked at the offshore wind farm. If the weather conditions turn out to be insufficient for the planned installation stage, the installation process is terminated (9) and reported to the supply chain partners to adjust the time schedule (10).

In connection with the provision of the components scheduled for pick-up, the respective main component manufacturer informs the logistics service provider about the readiness for dispatch (11, 12). Simultaneously, the operator of the construction vessel reports status information of the installation process, (13) that enables the logistics service provider to supply the components on schedule. Therefore, the beginning of every individual installation stage is reported by the operator of the construction vessel. The construction process of the foundation structure is subdivided into two steps. At first, the foundation structure is positioned and then the piles are rammed into the seabed to fix the foundation. For reasons of stability, the installation of the wind energy turbine is structured into three installation stages: lower tower section, other tower sections plus nacelle, and at last the set of rotor blades. Due to the average process time for each installation stage, the estimated time of arrival (ETA) of the construction vessel in the port for shipment can be passed on to the partners (14). It contributes to the on-schedule provision of the components by main component manufacturers and logistics service provider in the port of shipment.

The second part of the standardized information flow of the process concept is illustrated in Figure 8. If a severe damage of installation equipment or components occurs during the installation at sea that obstructs the work process, the installation is terminated, the new status is reported (15), and forwarded to the supply chain partners (16). Based on the ETA made available by the operator of the construction vessel (14) and the report of readiness for dispatch from the main component manufacturer (12) supplied by the web-based tool, the logistics service provider requests weather forecasts for the transport from the main component manufacturer to the port of shipment (17/18). This review of weather data is executed to ensure that a Just-In-Time (JIT) delivery of the components to the construction vessel is possible. If the weather conditions are sufficient for a JIT delivery, the components are supplied on barges alongside the construction vessel. If the weather conditions are insufficient for a JIT delivery, the components have to be stored temporarily in the port of shipment.

The logistics service provider schedules the transport of the individual components and reports it to the web-based tool that forwards the information to the respective main component manufacturer (19/20). The following pieces of status information are supplied by logistics service provider and construction vessel operator during operation. The operator of the construction vessel reports the completion of a foundation structure or an offshore wind energy turbine (21). The logistics service provider confirms the loading of a component at the main component manufacturer (22). This status information is provided on demand by the web-based tool.

When the construction vessel leaves the construction area to return to the port of shipment, then accordingly the status modification is reported (23) and forwarded to the logistics service provider (24). Based on this information, the arrival of the construction vessel can be retrieved for the exact hour and that ensures the supply of the scheduled components. The next status modifications are reported and forwarded when the construction vessel reaches the harbor entrance (25/26) and when the construction vessel reaches its landing and is ready for loading (27/28). After the loading of the components, it is confirmed by the operator of the construction vessel and the logistics service provider (29). This way, high additional costs due to false loadings can be prevented. Finally, the web-based tool forwards the loading confirmation to the respective main component manufacturer (30).

The supply of status information over the entire supply chain of the offshore wind energy industry by a collaborative web-based tool offers great potential to optimize the logistics processes of the supply chain partners. The communication effort of the whole process is immense due to the complexity of the installation process impeded by weather conditions. Therefore, the effort connected with the supply of status information has to be minimized to offer an improvement to current processes.

The idea is to automatically generate suitable information flows. The components of the wind energy turbines are usually transported in loading aids. These loading aids will be equipped with GPS (Global Positioning System) devices that allow the acquisition of positioning data. Furthermore, the components will be equipped with RFID (Radio Frequency Identification) devices to allow a distinct identification of the individual logistics object.

The information flow that is suitable for automation is described in the following. The first part of information flow that can be automated is the report of readiness for dispatch from the main component manufacturer (11). Due to the equipment of the components with GPS devices that provide position data and a distinct identification of the individual logistic object, the report of readiness for dispatch is automatically transmitted when the component reaches the predefined storage area of the individual main component manufacturer. The pieces of status information of the operator of the construction vessel (13) are also suitable for automation. The beginning of the installation of each individual component can be documented by the acquisition of the data from the installed RFID tags that allow a distinct identification and the documentation of the exact start time.

Another information flow that can be automated is the report of completion of foundation structures or complete wind energy turbines (21). This can be achieved in connection with other information flows. If the beginning of the installation is confirmed (13) and no termination of the installation (15) occurred, it can be concluded that an installation stage is finished when the vessel leaves the installation position. Further automated pieces of status information can be generated after the installation vessel leaves the construction site. Based on the GPS position of the vessel the entry of the harbor (25) and the arrival at the landing (27) can be concluded. The last automated information flow is the loading conformation by logistics service provider and vessel operator. The position of the components will be merged with the position of the construction vessel as soon as the component is loaded. The identification of the component will be conducted by reading the dedicated RFID tag.

3.4 Potential Transmission in Planning and Controlling System – Needs and Functions

Based on the design of the processes and the need to integrate the individual components into planning and controlling system, a link has to be created between the virtual world and the real one. All objects relevant to logistics (plant components, carriers, and lifting and handling gear) must be integrated into this system with their respective status. This makes it clear that such an instrument incorporates a monitoring system and the handling options of a Material Requirements Planning system. Due to the web-based, real-time-capable approach, monitoring and, consequently, the creation of information transparency across agents is possible.

The basic system components of such an operational planning and control instrument should offer the following functionalities in addition to a framework and project plan: localization, geofencing, status display, user concept, weather data, scenario analysis, interfaces for identification, location and communication technology, and further systems such as warning messages and reporting. It can be stated that the system only has to deal with the status and corresponding changes in the status. Only the information that has an effect on the logistics processes is relevant to the system.

4. Conclusion and Outlook

In this contribution, the authors presented a process concept for the offshore wind energy industry that results from a detailed process mapping of all supply chain partners. Based on these detailed processes, a process analysis was conducted and process weaknesses were identified. These process weaknesses were evaluated for their impact on the entire supply chain of the offshore wind energy industry. The analysis of the process weaknesses enabled us to derive and prioritize objectives for the development of the process concept. Finally, we developed a process concept that allows information transparency within the construction phase and its logistics processes. By offering standardized information and material flows over the entire logistics network as well as including a collaborative web-based tool as a central communication platform for the supply chain partners, the concept provides the basis for a reduction of operative costs, a substantial increase of process quality and shorter processing times for standardized operations under weather influence.

The process concept serves as a preparatory work for the development of a control stand for the offshore wind industry. Therefore, a web-based tool for the operative use within the logistics network of the offshore wind industry has to be developed. The operative web-based tool offers the possibility to monitor the current state of the installation process, assess the influences of forecasted and actual weather on the installation schedule, and to compare target and actual performance of the logistics network. The potential of the instrument that has been developed becomes clear when several wind farm projects are considered in parallel; this means that the complexity increases and each agent in the network has to act and respond based on the demands of the different projects. The instrument enables the dependency of each agent in the network to be illustrated in relation to the actions of the others. By equipping logistics objects with information and communication technology data of material, flows can be generated and the web-based tool can provide the actual position of individual components.

Acknowledgements

This contribution presents results of the cooperative project "Mon²Sea – real-time monitoring of the transport and handling of components for commissioning offshore wind turbines", which was supported by the German Federal Ministry for the Environment, Nature Conservation, and Nuclear Safety (BMU) as part of the 5th Energy Research Programme.

References

Ait Alla, A., Quandt, M., Beinke, T. and Freitag, M. (2016). Improving the decision-making process during the installation process of offshore wind farms by means of information sharing. pp. 144–150. *In*: Jin S. Chung, Muskulus, M., Kokkinis, T. and Alan M. Wang (eds.). Proceedings of the Twenty-sixth (2016) International Ocean and Polar Engineering Conference (ISOPE). Renewable Energy (Offshore Wind and Ocean) and Environment. California, USA.

Beinke, T., Freitag, M. and Zint, H.-P. (2015). Ressourcen-Sharing für eine bezahlbare Energiewende. Betrachtung der Produktions-und Errichtungslogistik der Offshore-Windenergie. Industrie 4.0 Mangament, 31(4): S. 7–11.

Bodenstab, M., Sagert, Ch. and Beinke, T. (2014). Konzeption einer operativen Planung und Steuerung für die Offshore-Windenergielogistik der Errichtungsphase. *In*: Thoben, K.-D., Haasis, H.-D. and Lewandowski, M. (Hrsg.): Logistik für die Windenergie. Herausforderungen und Lösungen für moderne Windkraftwerke, epubli GmbH, Berlin, S. 75–84.

Briese, D. (2016). Ausblick: Trends und Entwicklungen, Chancen und Risiken. *In*: Jessica Wegener, Lina Harms, Marie Hartinger und Andreas Findeisen (Hg.): Schnittstellenmanagement Offshore Wind. Praxishandbuch. S. 281–288. Hamburg, Germany.

Briese, D. and Westhäuser, M. (2013). Zukunftsperspektiven und Herausforderungen der Offshore-Windenergie. *In*: Jörg Böttcher (Hg.): Handbuch Offshore-Windenergie. Rechtliche, technische und wirtschaftliche Aspekte. S. 1–54. Oldenbourg Verlag. München, Germany.

Burckhardt, C. (2013). Logistik- und Wartungskonzepte. In: Jörg Böttcher (Hg.): Handbuch Offshore-Windenergie. Rechtliche, technische und wirtschaftliche Aspekte. S. 423–431. Oldenbourg Verlag. München, Germany.

Federal Ministry for Economic Affairs and Energy (2015). Offshore wind energy. An overview of activities in Germany. Berlin. Accessed 12.09.2017. http://www.bmwi.de/Redaktion/EN/Publikationen/offshore-wind-energy.pdf?__blob=publicationFile&v=3.

Ferrer, J., MacCawley, A., Maturana, S., Toloza, S. and Vera, J. (2008). An optimization approach for scheduling wine grape harvest operations. International Journal of Production Economics, (112): 985–999.

Gasch, R. and Twele, J. (2013). Windkraftanlagen. Grundlagen, Entwurf, Planung und Betrieb. Vieweg + Teubner. Wiesbaden, Germany.

German Federal Government (2010). Energiekonzept. für eine umweltschonende, zuverlässige und bezahlbare Energieversorgung. Accessed 12.09.2017. http://www.bundesregierung. de/ContentArchiv/DE/Archiv17/_Anlagen/2012/02/energiekonzept-final.pdf?__ blob=publicationFile&v=5.

Görges, Michael, Möller, Jörg and Shao, Jincheng. (2014). Simulationsgestützte Planung und Steuerung in der Offshore-Logistik. *In*: Klaus-Dieter Thoben, Hans-Dietrich Hassis und Marco Lewandowski (Hg.): Logistik für die Windenergie. Herausforderungen und Lösungen für moder-ne Windkraftwerke. Berlin: epubli, S. 49–58.

Gille, Denny. (2011). Kein Standard in Sicht. *In*: Erneuerbare Energien, 21(1): S. 38–43.

Gudehus, T. and Kotzab, H. (2012). Comprehensive Logistics. Springer, Berlin Heidelberg.

Hau, E. (2014). Windkraftanlagen. Grundlagen, Technik, Einsatz, Wirtschaftlichkeit. Springer. Berlin, Germany.

Heidmann, R. (2010). Anfänge im Hinterland. *In*: Erneuerbare Energien, (20). S. 50–53.

Heidmann, R. (2015). Windenergie und Logistik. Losgröße 1: Logistikmanagement im Maschinen- und Anlagenbau mit geringen Losgrößen. Beuth. Berlin, Germany.

Hobohm, J., Krampe, L., Peter, F., Gerken, A., Heinrich, P. and Richter, M. (2013). Kostensenkungspotenziale der Offshore-Windenergie in Deutschland. Langfassung. Berlin. Accessed 12.09.2017. https://www.prognos.com/uploads/tx_atwpubdb/130822_Prognos_ Fichtner_Studie_Offshore-Wind_Lang_de.pdf.

Ishikawa, K. (1986). Guide to Quality Control. Asian Productivity Organization, Tokyo.

Klaus, P., Krieger, W. and Krupp, M. (2012). Gabler Lexikon Logistik. Management logistischer Netzwerke und Flüsse. 5. ed. Gabler Verlag. Wiesbaden, Germany.

Kovácsa, A., Erdősa, G., János Viharosa, Z. and Monostori, L. (2011). A system for the detailed scheduling of wind farm. CIRP Annals - Manufacturing Technology, 60(1): 497–501.

Le Gal, P., Le Masson, J., Bezuidenhout, C. and Lagrange, L. (2009). Coupled modelling of sugarcane supply planning and logistics as a management tool. Computers and Electronics in Agriculture, 68: 168–177.

Lee, S., Pena-Mora, F. and Park, M. (2006). Dynamic planning and control methodology for strategic and operational construction project management. Automation in Construction, 15: 84–97.

Lütjen, M. and Karimi, H.R. (2012). Approach of a port inventory control system for the offshore installation of wind turbines. Proceedings of the Twenty-second (2012) International Offshore and Polar Engineering Conference (ISOPE). pp. 502–508. Rhodes, Greek.

McCrea, M.V., Sherali, H.D. and Trani, A.A. (2008). A probabilistic framework for weather-based rerouting and delay estimations within an Airspace Planning model. Transportation Research Part C: Emerging Technologies, 16: 410–431.

Moccia, J., Arapogianni, A., Williams, D., Philips, J. and Hassan, G. (2011). Wind in our Sails – The coming of Europe's offshore wind energy industry. The European Wind Energy Association.

Oelker, S., Ait Alla, A., Lewandowski, M. and Freitag, M. (2016). Planning of maintenance resources for the service of offshore wind turbines by means of simulation. pp. 303–312. *In*: Proceedings of the 5th International Conference on Dynamics in Logistics (LDIC 2016), Springer.

Quandt, M., Beinke, T., Ait Alla, A. and Freitag, M. (2017). Simulation based investigation of the impact of information sharing on the offshore wind farm installation process. *In*: Journal of Renewable Energy, 1/2017, pp. 11.

Reichert, F., Kunze, R. and Kitvarametha, S. (2012). Expedition Offshore Windlogistik. Hintergrundinformation. München, Germany.

Rohrig, K., Richts, Ch., Bofinger, S., Jansen, M., Seifert, M., Pfaffel, S. and Durstewitz, M. (2013). Energiewirtschaftliche Bedeutung der Offshore-Windenergie für die Energiewende. Fraunhofer-Institut für Windenergie und Energieysteme. Kassel. Accessed 12.09.2017. https://www.offshore-stiftung.de/sites/offshorelink.de/files/documents/SOW_Download_Langfassung-Energiewirtsch aftlicheBedeutungderOffshore-Windenergie.pdf.

Sankar, N. and Prabhu, B. (2001). Modified approach for prioritization of failures in a system failure mode and effect analysis. International Journal of Quality & Reliability Management, 18: 324–335.

Scholz-Reiter, B., Heger, J., Lütjen, M. and Schweizer, A. (2011). A MILP for installation scheduling of offshore wind farms. International Journal of Mathematical Models and Methods in Applied Sciences, 5(1): 371–378.

Schulte, Ch. (2013). Logistik. Wege zur Optimierung der Supply Chain. 6. ed. Vahlen. München, Germany.

Slack, N., Chambers, S. and Johnston, R. (2010). Operations Management. Pearson Education Ltd, Essex.

Stohlmeyer, H. and Ondraczek, J. (2013). Darstellung und Mitigierung zentraler Fertigungsrisiken. Jörg Böttcher (Hg.): Handbuch Offshore-Windenergie. Rechtliche, technische und wirtschaftliche Aspekte. S. 330–352. Oldenbourg Verlag. München, Germany.

Swamidass, P. (ed.) (2000). Encyclopedia of production and manufacturing management. Boston Dordrecht London. Kluwer Academic Publishers.

Thoben, K.-D., Hassis, H.-D. and Lewandowski, M. (2014). Offshore-Logistik - Herausforderungen an die Logistik für die Windenergie und Potentiale der angewandten Forschung. Thoben, K.-D., Hassis, H.-D., Lewandowski, M. (Hg.): Logistik für die Windenergie. Herausforderungen und Lösungen für moderne Windkraftwerke. S. 3–11. Epubli. Berlin, Germany.

Tyapin, I., Hovland, G. and Jorde, J. (2011). Comparison of Markov Theory and Monte Carlo Simulations for Analysis of Marine Operations Related to Installation of an Offshore Wind Turbine. Proceedings of the 24th International Congress on Condition Monitoring (COMADEM). Stavanger, Norway, pp. 1071–1081.

Zhang, J., Chowdhury, S., Tong, W. and Messac, A. (2012). Optimal Preventive Maintenance Time Windows for Offshore Wind Farms Subject to Wake Losses. Proceedings of the 14th AIAA/ ISSMO Multidisciplinary Analysis and Optimization Conference, Indianapolis, USA, pp. 1–13.

3

Estimation-Based Ocean Flow Field Reconstruction Using Profiling Floats

Huazhen Fang,[1,]* *Raymond A. de Callafon*[2] *and* *Jorge Cortés*[2]

1. Introduction

This chapter presents a study of ocean flow field reconstruction using profiling floats. Buoyancy-controlled, semi-autonomous profiling floats provide an economic and flexible means to monitor ocean flows. However, a key challenge in enabling this capability lies in how to extract the velocities of the ocean flow at sampled locations from the recorded motion data of floats. This leads to a problem that can be cast as the joint estimation of a nonlinear dynamic system's inputs and states. This chapter develops solutions from the perspective of Bayesian estimation and evaluates their application to ocean flow field reconstruction through simulations. The proposed results may also be useful in many other scientific and engineering problems involving input and state estimation.

---

[1] Department of Mechanical Engineering, University of Kansas, Lawrence, KS 66045, USA.
[2] Department of Mechanical and Aerospace Engineering, University of California, San Diego, CA 92093, USA.
* Corresponding author

1.1 Background

The oceans, which cover over two-thirds of the Earth's surface, have been an essential part of human life as food sources and transportation routes for thousands of years. Today, the world is increasingly looking to them in seeking solutions to various grand challenges such as natural resource and energy shortage and climate change (Costanza 1999). Associated with this trend is a growing commitment from the research community to understand the oceans with greater spatial and temporal coverage and finer accuracy. A fundamental research problem is to monitor ocean flows that result from continuous, directed movement of ocean water and that significantly impact the marine environment, maritime transport, pollution spread, and global climate.

Recent years have witnessed an exponential interest in deploying autonomous underwater vehicles (AUVs) for flow monitoring which are capable of operating in dangerous underwater environments without the need for human presence. Crucial for the development of advanced AUVs is an integration of control, communication, and computation technologies; which has benefited significantly from the recent sweeping advances in these fields. Technological sophistication, however, comes initially with high economic costs, making AUVs expensive and not widely available to oceanographic research. This has provided impetus for using small inexpensive AUVs known as profiling floats. A profiling float only has buoyancy control to change its vertical position and laterally drifts along the flow. Despite underactuation, it can travel across the ocean at different depths for long durations to observe temperature, salinity, and currents. About 3,900 such instruments, through the internationally collaborative Argo program, have been deployed in global oceans to collect data for climate and oceanographic research (Riser et al. 2016).

Our work here considers a profiling-float-based ocean observing system (Colgan 2006, Ouimet and Cortés 2014), which is schematically illustrated in Figure 1a. This system consists of a few profiling floats. They can be released at different locations in a region of interest in the ocean. Then, each one travels in cycling movement patterns of submerging/surfacing (see Figure 1b) with lateral motion driven by ocean currents and vertical motion regulated by buoyancy. While underwater, each float stores a time record of its current depth, acceleration, and other oceanographically relevant quantities. When coming up to the surface, it determines its geographical location and then transmits all the data *via* a communication satellite for analysis and computation. The acceleration is continually measured by an on-board accelerometer and the position by a satellite-based Global Positioning System (GPS). It should be noted that a float's position is intermittently available—it can be measured only when the float is on the surface because GPS signals are seriously attenuated while underwater.

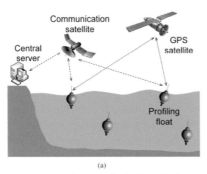
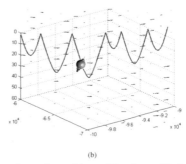

(a)   (b)

Figure 1. (a) The scenario for flow field reconstruction based on submersible profiling floats; (b) the traveling profile (submerging/surfacing) of a profiling float.

The objective here is to reconstruct the flow field and monitor the movement of the profiling floats. Hence, it is necessary to simultaneously estimate the flow velocities, which act as external inputs applied to a float, and the float's underwater positions and velocities, which make up the float's states, using the float's position and acceleration measurements. This leads us to investigate the problem of simultaneous input and state estimation (SISE) for a nonlinear system based on the system's output measurement data. It is noteworthy that problems of a similar kind also arise in other fields with examples including fault detection, disturbance rejection, vehicle tire-road friction estimation, and weather forecasting (Schubert et al. 2012, Imine et al. 2006, Kitanidist 1987), where a system's input and state variables are both unknown and need to be estimated. This observation, together with the problem of flow field reconstruction specifically considered here, motivates us to develop general-purpose solutions.

1.2 Literature Survey

Since the 1960s, the systems and control community has explored state estimation with unknown inputs. The earliest relevant work, to our knowledge, is (Friedland 1969), which studies state estimation with a constant but unknown disturbance input. The proposed solution is to augment the state vector to include the disturbance and then apply the Kalman filter (KF). In (Kitanidis 1987), minimum variance unbiased estimation (MVUE) is exploited to estimate states in the presence of completely unknown inputs. Based on (Kitanidis 1987), a series of MVUE-based methods are proposed in (Darouach and Zasadzinski 1997, Darouach et al. 2003, Cheng et al. 2009), along with a detailed analysis of the existence and stability of these estimators. For these studies, the focus is on enabling state estimation despite unknown inputs, leaving aside input estimation. However, the unknown inputs are also important for many real-world systems and, if successfully estimated, can be useful for system analysis and control synthesis. As such, SISE presents even more appeal and has attracted significant attention from researchers.

The literature on SISE includes two main subjects: simultaneous input and state filtering (SISF) and smoothing (SISS). As the names suggest, SISF seeks to estimate the input and state at the current time instant using the history of output measurements, and SISS seeks to estimate the input and state at a past time instant using measurements inclusive of those lagging behind that time instant. The former thus allows for real-time estimation, and the latter requires off-line computation leading to better estimation accuracy in spite of an increase in estimator complexity.

For SISF, an early contribution is (Mendel 1977), which adapts the KF to estimate a linear system's states along with covariance-known white noise disturbances. More recent works usually consider completely unknown inputs and develop SISF techniques by modifying some existing state estimation approaches. Among them, we highlight those based on the KF (Hsieh 2000, Hsieh 2010, Hsieh 2011), moving horizon estimation (MHE) (Pina and Botto 2006), $\mathcal{H}_\infty$-filtering (You et al. 2008), sliding mode observers (Floquet et al. 2007), and MVUE (Gillijns and De Moor 2007a, Gillijns and De Moor 2007b, Fang et al. 2008, Fang et al. 2011, Fang and de Callafon 2012, Yong et al. 2016, Shi et al. 2016). Though reaching a certain level of maturity for linear systems, SISF can become rather complicated when nonlinear systems are considered. In (Corless and Tu 1998, Ha and Trinh 2004), SISF methods are developed for a special class of nonlinear deterministic systems which consist of a nominally linear part and a nonlinear part. The work (Hsieh 2013) decouples the unknown inputs from the nonlinear system and then extends linear SISF methods to handle the estimation. In our prior work (Fang et al. 2013, Fang and de Callafon 2011, Fang et al. 2017), we have shown that Bayesian estimation is a viable approach to cope with SISF for generic-form nonlinear stochastic systems from low to high dimensions.

When it comes to SISS, and despite its significance, only few studies have been reported. Extrapolating the Bayesian approach in (Fang and de Callafon 2011, Fang et al. 2013), we proposed SISS algorithms for nonlinear systems in (Fang and de Callafon 2013, Fang et al. 2015) and also specialized them to linear systems. A linear smoothing algorithm is also developed in (Yong et al. 2014) as an extension of MVUE-based SISF in (Yong et al. 2016). An exhaustive search shows no more results in the literature other than these studies. It is worth pointing out that ocean flow field estimation is a problem well-suited to SISS because a typical ocean flow or circulation changes on time scales ranging from a few days to a season, thus allowing off-line but more accurate estimation.

1.3 Overview of this Chapter

This chapter summarizes our previous work (Fang and de Callafon 2011, Fang et al. 2013, Fang and de Callafon 2013, Fang et al. 2015) to present a systematic introduction of Bayesian-estimation-based SISF and SISS methods for nonlinear systems. The core idea of Bayesian estimation is updating the probabilistic belief of an unknown variable using measurement data and then extracting the best estimate in a probabilistic sense. This methodology has been proven useful for constructing various state and parameter estimation techniques over the past decades. Our study further advances this actively researched area to investigate SISE. We start by developing Bayesian SISF and SISS paradigms to conceptually look at and understand the SISE problems from the perspective of statistical estimation. This view opens up the possibility of developing Bayesian SISF and SISS estimators. Building upon the Bayesian paradigm, we formulate maximum a posteriori probability (MAP) estimation problems and solve them using the Gauss-Newton method which is a numerical optimization approach capable of overcoming the effects of a system's intrinsic nonlinearity on estimation accuracy. For each case of SISF and SISS, we develop separate investigations for systems with and without direct input-to-output feedthrough, leading to a set of estimation algorithms custom-built according to the system structure. We conduct a simulation study to apply the proposed algorithms in the reconstruction of an ocean flow field using observation data collected by a group of buoyancy-controlled profiling floats. The estimation performance validates the effectiveness of the proposed algorithms.

2. SISF for Systems with Direct Feedthrough

This section studies the problems of SISF for dynamic systems with direct feedthrough. It starts with developing a Bayesian estimation principle and then derives an SISF algorithm using the idea of Bayesian MAP estimation.

2.1 The Bayesian Paradigm

Consider a nonlinear system with direct input-to-output feedthrough:

$$\begin{cases} \mathbf{x}_{k+1} = \mathbf{f}(\mathbf{x}_k, \mathbf{u}_k) + \mathbf{w}_k \\ \mathbf{y}_k = \mathbf{h}(\mathbf{x}_k, \mathbf{u}_k) + \mathbf{v}_k \end{cases}, \tag{1}$$

where $\mathbf{x} \in \mathbb{R}^n$ is the state vector, $\mathbf{u} \in \mathbb{R}^m$ the input vector, $\mathbf{y} \in \mathbb{R}^p$ the measurement vector, and $\mathbf{w} \in \mathbb{R}^n$ and $\mathbf{v} \in \mathbb{R}^m$ mutually independent zero-mean white Gaussian noise sequences, with covariances $\mathbf{Q}_k$ and $\mathbf{R}_k$, respectively. The mappings $\mathbf{f}: \mathbb{R}^n \times \mathbb{R}^m \to \mathbb{R}^n$ and $\mathbf{h}: \mathbb{R}^n \times \mathbb{R}^m \to \mathbb{R}^p$ are the state transition and measurement functions, respectively, which are assumed to be C^1. We also assume $\nabla_u \mathbf{h}$ has full rank. For the above system, our objective is to estimate $\mathbf{u}_k$ and $\mathbf{x}_k$ based on the measurement set $\mathbf{Y}_{1:k} = \{\mathbf{y}_1, \mathbf{y}_2, \ldots, \mathbf{y}_k\}$. To build a Bayesian estimator, one is interested in finding out the probability density functions (pdf's) of $\mathbf{u}_k$ and $\mathbf{x}_k$ conditioned on $\mathbf{Y}_{1:k}$, i.e., $p(\mathbf{u}_k, \mathbf{x}_k|\mathbf{y}_{1:k})$. As it is also desirable to achieve sequential estimation, the problem becomes how to enable the passing to $p(\mathbf{u}_k, \mathbf{x}_k|\mathbf{y}_{1:k})$ from $p(\mathbf{u}_{k-1}, \mathbf{x}_{k-1}|\mathbf{y}_{1:k-1})$. Akin to Bayesian state estimation (Candy 2009), this can be accomplished in a two-step procedure of prediction and update.

The step of prediction is used to determine the conditional pdf $p(\mathbf{x}_k|\mathbf{y}_{1:k-1})$. By the Chapman-Kolmogorov equation (Honerkamp 1993), we have

$$p(\mathbf{x}_k|\mathbf{Y}_{1:k-1}) = \iint p(\mathbf{x}_k|\mathbf{u}_{k-1}, \mathbf{x}_{k-1}, \mathbf{Y}_{1:k-1}) \cdot p(\mathbf{u}_{k-1}, \mathbf{x}_{k-1}|\mathbf{Y}_{1:k-1})\, d\mathbf{u}_{k-1} d\mathbf{x}_{k-1}.$$

We note that $p(\mathbf{x}_k|\mathbf{u}_{k-1}, \mathbf{x}_{k-1}, \mathbf{Y}_{1:k-1}) = p(\mathbf{x}_k|\mathbf{u}_{k-1}, \mathbf{x}_{k-1})$, since $\mathbf{x}_k$ depends on only $\mathbf{u}_{k-1}$ and $\mathbf{x}_{k-1}$ because of the Markovian state propagation as shown in (1). Hence, it follows that

$$p(\mathbf{x}_k|\mathbf{Y}_{1:k-1}) = \iint p(\mathbf{x}_k|\mathbf{u}_{k-1}, \mathbf{x}_{k-1}) \cdot p(\mathbf{u}_{k-1}, \mathbf{x}_{k-1}|\mathbf{Y}_{1:k-1})d\mathbf{u}_{k-1}d\mathbf{x}_{k-1}. \qquad (2)$$

When the measurement $\mathbf{y}_k$ arrives, it can be used to update $p(\mathbf{x}_k|\mathbf{Y}_{1:k-1})$ along with the conditional pdf of $\mathbf{u}_k$ (because it is the first measurement conveying information about $\mathbf{u}_k$) *via* determining $p(\mathbf{u}_k, \mathbf{x}_k|\mathbf{Y}_{1:k})$. To proceed, we make the following assumption:

(A1)$\{\mathbf{u}_k\}$ is a white process, independent of $\mathbf{x}_0$, $\{\mathbf{w}_k\}$, and $\{\mathbf{v}_k\}$.

Here, 'white' means that $\mathbf{u}_k$ and $\mathbf{u}_l$ are independent random vectors for $k \neq l$. Such a whiteness assumption is inspired by (Robinson 1957), which has been a foundation for many seismic data processing algorithms. The intuitions underlying it are: (1) $\mathbf{u}_k$, completely unknown to us, may assume all possible values; (2) from the knowledge of $\mathbf{u}_k$ we cannot predict $\mathbf{u}_l$ for $k \neq l$. A similar treatment of $\{\mathbf{u}_k\}$ as a stochastic process is proposed in (Friedland 1969), which yet assumes the wide-sense description of $\mathbf{u}_k$ as known. By (A1), $\mathbf{u}_k$ is independent of $\mathbf{x}_k$ and $\mathbf{Y}_{1:k-1}$ (Gut 2005, Theorem 10.4, pp. 71).

Using the Bayes' rule repeatedly, we obtain

$$p(\mathbf{u}_k, \mathbf{x}_k|\mathbf{Y}_{1:k}) = \frac{p(\mathbf{y}_k|\mathbf{u}_k,\mathbf{x}_k,\mathbf{Y}_{1:k-1}) \cdot p(\mathbf{u}_k,\mathbf{x}_k|\mathbf{Y}_{1:k-1})}{p(\mathbf{y}_k|\mathbf{Y}_{1:k-1})}.$$

Note that $p(\mathbf{y}_k|\mathbf{u}_k, \mathbf{x}_k, \mathbf{Y}_{1:k-1}) = p(\mathbf{y}_k|\mathbf{u}_k, \mathbf{x}_k)$ due to the fact that $\mathbf{y}_k$ entirely depends on $\mathbf{u}_k$ and $\mathbf{x}_k$, and that $p(\mathbf{u}_k, \mathbf{x}_k|\mathbf{Y}_{1:k-1}) = p(\mathbf{x}_k|\mathbf{Y}_{1:k-1}) \cdot p(\mathbf{u}_k)$ as a result of $\mathbf{u}_k$'s independence from $\mathbf{x}_k$ and $\mathbf{Y}_{1:k-1}$. Consequently,

$$p(\mathbf{u}_k, \mathbf{x}_k | \mathbf{Y}_{1:k}) = \frac{p(\mathbf{y}_k | \mathbf{u}_k, \mathbf{x}_k) \cdot p(\mathbf{x}_k | \mathbf{Y}_{1:k-1}) \cdot p(\mathbf{u}_k)}{p(\mathbf{y}_k | \mathbf{Y}_{1:k-1})}$$

One can see that $p(\mathbf{u}_k)/p(\mathbf{y}_k | \mathbf{Y}_{1:k-1})$ plays the role of a proportionality coefficient. This implies

$$p(\mathbf{u}_k, \mathbf{x}_k | \mathbf{Y}_{1:k}) \propto p(\mathbf{y}_k | \mathbf{u}_k, \mathbf{x}_k) \cdot p(\mathbf{x}_k | \mathbf{Y}_{1:k-1}). \tag{3}$$

Here, (2) and (3) form the Bayesian SISF paradigm for systems with direct feedthrough. Sequentially updating them not only provides a conceptual Bayesian solution to the considered SISF problem but also yields a statistical framework within which different SISE methods can be developed. Our next step is to derive an SISF algorithm by formulating and solving an MAP estimation problem based on the proposed Bayesian paradigm.

2.2 SISF Algorithm Development

Let us begin with some Gaussian distribution assumptions for concerned pdf's. Specifically, we assume

$$(A2)\, p(\mathbf{u}_k, \mathbf{x}_k | \mathbf{Y}_{1:k}) \sim \mathcal{N}\left(\begin{bmatrix} \hat{\mathbf{u}}_{k|k} \\ \hat{\mathbf{x}}_{k|k} \end{bmatrix}, \begin{bmatrix} \mathbf{P}^{u}_{k|k} & \mathbf{P}^{ux}_{k|k} \\ (\mathbf{P}^{ux}_{k|k})^T & \mathbf{P}^{x}_{k|k} \end{bmatrix}\right),$$

$$(A3)\, p(\mathbf{y}_k | \mathbf{u}_k, \mathbf{x}_k) \sim \mathcal{N}(\mathbf{h}(\mathbf{u}_k, \mathbf{x}_k), \mathbf{R}_k),$$

$$(A3)\, p(\mathbf{x}_k | \mathbf{Y}_{1:k-1}) \sim \mathcal{N}(\hat{\mathbf{x}}_{k|k-1}, \mathbf{P}^{x}_{k|k-1}),$$

where $\hat{\mathbf{u}}_{k|k}$ is the estimate of $\mathbf{u}_k$ given $\mathbf{Y}_k$ with associated covariance $\mathbf{P}^{u}_{k|k}$, $\hat{\mathbf{x}}_{k|k-1}$ and $\hat{\mathbf{x}}_{k|k}$ are the estimates of $\mathbf{x}_k$ given $\mathbf{Y}_{1:k-1}$ and $\mathbf{Y}_{1:k}$ with covariances $\mathbf{P}^{x}_{k|k-1}$ and $\mathbf{P}^{x}_{k|k}$, respectively. Ideally, if knowledge of $p(\mathbf{u}_k, \mathbf{x}_k | \mathbf{Y}_{1:k})$ is available for each k, $\hat{\mathbf{u}}_{k|k}$ and $\hat{\mathbf{x}}_{k|k}$ can be readily obtained by MAP or some other way. However, contrary to this ideal, determining $p(\mathbf{u}_k, \mathbf{x}_k | \mathbf{Y}_{1:k})$ accurately is known as an intractable issue for nonlinear systems. In order to overcome this problem, (A2)–(A4) are made to approximately describe the pdf's by replacing each with a Gaussian distribution with the same mean and covariance. Assumptions on Gaussian distributions analogous to (A2)–(A4) are commonly held in nonlinear estimation algorithms, e.g. (Anderson and Moore 1979, Bell and Cathey 1993, Spinello and Stilwell 2010).

Based on the idea of MAP estimation, we intend to enable state prediction by considering

$$\hat{\mathbf{x}}_{k|k-1} = \underset{\mathbf{x}_k}{\operatorname{argmax}}\, p(\mathbf{x}_k | \mathbf{Y}_{1:k-1}) \tag{4}$$

which maximizes the probabilistic presence of $\mathbf{x}_k$ given $\mathbf{Y}_{1:k-1}$. To solve this problem, we first look at the first-order Taylor series expansion of $\mathbf{f}(\mathbf{u}_k, \mathbf{x}_k)$ around $(\hat{\mathbf{u}}_{k|k}, \hat{\mathbf{x}}_{k|k})$:

$$\mathbf{f}(\mathbf{u}_k, \mathbf{x}_k) \approx \mathbf{f}(\hat{\mathbf{u}}_{k|k}, \hat{\mathbf{x}}_{k|k}) + \nabla \mathbf{f}(\hat{\mathbf{u}}_{k|k}, \hat{\mathbf{x}}_{k|k}) \begin{bmatrix} \mathbf{u}_k - \hat{\mathbf{u}}_{k|k} \\ \mathbf{x}_k - \hat{\mathbf{x}}_{k|k} \end{bmatrix}, \tag{5}$$

where $\nabla \mathbf{f} = [\nabla_\mathbf{u} \mathbf{f} \quad \nabla_\mathbf{x} \mathbf{f}]$. Then by (A2), (2) and (5), the approximate solution to (4) is given by

$$\hat{\mathbf{x}}_{k|k-1} = \mathbf{f}(\hat{\mathbf{u}}_{k-1|k-1}, \hat{\mathbf{x}}_{k-1|k-1}), \tag{6}$$

with the associated prediction error covariance $\mathbf{P}^\mathbf{x}_{k|k-1}$ given by

$$\mathbf{P}^\mathbf{x}_{k|k-1} \approx \nabla \mathbf{f}(\hat{\mathbf{u}}_{k-1|k-1}, \hat{\mathbf{x}}_{k-1|k-1}) \begin{bmatrix} \mathbf{P}^\mathbf{u}_{k-1|k-1} & \mathbf{P}^\mathbf{ux}_{k-1|k-1} \\ (\mathbf{P}^\mathbf{ux}_{k-1|k-1})^T & \mathbf{P}^\mathbf{x}_{k-1|k-1} \end{bmatrix} \nabla \mathbf{f}^T(\hat{\mathbf{u}}_{k-1|k-1}, \hat{\mathbf{x}}_{k-1|k-1}) + \mathbf{Q}_{k-1}. \tag{7}$$

Then, (6) and (7) constitute the prediction formulae together, computing the state prediction and prediction error covariance, respectively.

Let us now consider updating $p(\mathbf{x}_k|\mathbf{Y}_{1:k-1})$ using $\mathbf{y}_k$ and define the following MAP estimator:

$$\begin{bmatrix} \hat{\mathbf{u}}_{k|k} \\ \hat{\mathbf{x}}_{k|k} \end{bmatrix} = \underset{\mathbf{u}_k, \mathbf{x}_k}{\mathrm{argmax}} \, p(\mathbf{u}_k, \mathbf{x}_k|\mathbf{Y}_{1:k}). \tag{8}$$

We further define the MAP cost function as $L(\mathbf{u}_k, \mathbf{x}_k) = p(\mathbf{u}_k, \mathbf{x}_k|\mathbf{Y}_{1:k})$. According to (3) and (A3)–(A4), one has

$$L(\mathbf{u}_k, \mathbf{x}_k) = \lambda \cdot \exp[-\boldsymbol{\alpha}_k^T \mathbf{R}_k^{-1} \boldsymbol{\alpha}_k - \boldsymbol{\beta}_k^T (\mathbf{P}^\mathbf{x}_{k|k-1})^{-1} \boldsymbol{\beta}_k],$$

where λ combines all the constants, $\boldsymbol{\alpha}_k = \mathbf{y}_k - \mathbf{h}(\mathbf{u}_k, \mathbf{x}_k)$ and $\boldsymbol{\beta}_k = \mathbf{x}_k - \hat{\mathbf{x}}_{k|k-1}$. It is easier to deal with the logarithmic cost function $\ell(\mathbf{u}_k, \mathbf{x}_k) = -\ln L(\mathbf{u}_k, \mathbf{x}_k)$:

$$\ell(\mathbf{u}_k, \mathbf{x}_k) = \delta + \mathbf{r}^T(\mathbf{u}_k, \mathbf{x}_k) \cdot \mathbf{r}(\mathbf{u}_k, \mathbf{x}_k), \tag{9}$$

where $\delta = -\ln \lambda$ and

$$\mathbf{r}(\mathbf{u}_k, \mathbf{x}_k) = \begin{bmatrix} \mathbf{R}_k^{-\frac{1}{2}} \boldsymbol{\alpha}_k \\ (\mathbf{P}^\mathbf{x}_{k|k-1})^{-\frac{1}{2}} \boldsymbol{\beta}_k \end{bmatrix}.$$

Thus, (8) can be equivalently written as

$$\begin{bmatrix} \hat{\mathbf{u}}_{k|k} \\ \hat{\mathbf{x}}_{k|k} \end{bmatrix} = \underset{\mathbf{u}_k, \mathbf{x}_k}{\mathrm{argmin}} \, \ell(\mathbf{u}_k, \mathbf{x}_k). \tag{10}$$

The MAP optimization in (10) usually defies the development of a closed-form solution when considered for a nonlinear system. However, as a nonlinear least-squares problem, it can be numerically addressed using the Gauss-Newton method (Björck 1996). The classical Gauss-Newton method can iteratively compute the sequences of approximations $\hat{\mathbf{u}}_k^{(i)}$ and $\hat{\mathbf{x}}_k^{(i)}$, where (i) denotes the iteration step. Specifically,

$$\hat{\boldsymbol{\xi}}_k^{(i+1)} = \hat{\boldsymbol{\xi}}_k^{(i)} - [\nabla_{\xi}^T \mathbf{r}(\hat{\boldsymbol{\xi}}_k^{(i)}) \cdot \nabla_{\xi} \mathbf{r}(\hat{\boldsymbol{\xi}}_k^{(i)})]^{-1} \cdot \nabla_{\xi}^T \mathbf{r}(\hat{\boldsymbol{\xi}}_k^{(i)}) \cdot \mathbf{r}(\hat{\boldsymbol{\xi}}_k^{(i)}), \tag{11}$$

where $\boldsymbol{\xi}_k = [\mathbf{u}_k^T \ \mathbf{x}_k^T]^T$, and $\nabla_{\xi} \mathbf{r} = [\nabla_{\mathbf{u}} \mathbf{r} \ \nabla_{\mathbf{x}} \mathbf{r}]$. One can let the initial guess be $\hat{\boldsymbol{\xi}}_k^{(0)} = [\mathbf{0}^T \ \hat{\mathbf{x}}_{k|k-1}^T]^T$ for convenience, though it can be set to arbitrary values. The iteration continues until the iteration step (i) reaches the preselected maximum i_{max} or the difference between two consecutive iterations is less than a preselected small value. Then, $\hat{\boldsymbol{\xi}}_k^{(i)}$ obtained in the final iteration will be exported and assigned to $\hat{\mathbf{u}}_{k|k}$ and $\hat{\mathbf{x}}_{k|k}$, respectively. The iteration process in (11) refines the input and state estimates continually by re-evaluating the joint estimator around the latest estimated input and state operating point. Despite demanding more computational power, the iterative refinement enhances not only the estimation performance but also the robustness to nonlinearities. In its practical use, one can try to strike a balance between computational complexity and estimation performance by selecting a proper stopping condition.

The estimation error covariance is equal to the inverse of the Fisher information matrix, as is known for MAP estimators under Gaussian distributions (Mutambara 1998). Then, we have

$$\begin{bmatrix} \mathbf{P}_{k|k}^{u} & \mathbf{P}_{k|k}^{ux} \\ (\mathbf{P}_{k|k}^{ux})^T & \mathbf{P}_{k|k}^{x} \end{bmatrix} = \mathcal{F}^{-1}(\hat{\mathbf{u}}_{k|k}, \hat{\mathbf{x}}_{k|k}), \tag{12}$$

where $\mathcal{F}$ is the Fisher information matrix defined as

$$\mathcal{F} = \begin{bmatrix} \mathcal{F}^{u} & \mathcal{F}^{ux} \\ (\mathcal{F}^{ux})^T & \mathcal{F}^{x} \end{bmatrix} = \mathrm{E}\left(\begin{bmatrix} \nabla_{\mathbf{u}}^T \ell \\ \nabla_{\mathbf{x}}^T \ell \end{bmatrix} [\nabla_{\mathbf{u}} \ell \ \nabla_{\mathbf{x}} \ell] \right). \tag{13}$$

The explicit formulae for the involved gradients are as follows:

$$\nabla_{\mathbf{u}} \mathbf{r} = \begin{bmatrix} -\mathbf{R}^{-\frac{1}{2}} \nabla_{\mathbf{u}} \mathbf{h} \\ \mathbf{0} \end{bmatrix}, \ \nabla_{\mathbf{x}} \mathbf{r} = \begin{bmatrix} -\mathbf{R}^{-\frac{1}{2}} \nabla_{\mathbf{x}} \mathbf{h} \\ (\mathbf{P}_{k|k-1}^{x})^{-\frac{1}{2}} \end{bmatrix},$$

$$\nabla_{\mathbf{u}} \ell = \mathbf{r}^T \nabla_{\mathbf{u}} \mathbf{r} = \boldsymbol{\alpha}^T \mathbf{R}^{-1} \nabla_{\mathbf{u}} \mathbf{h},$$

$$\nabla_{\mathbf{x}} \ell = \mathbf{r}^T \nabla_{\mathbf{x}} \mathbf{r} = \boldsymbol{\alpha}^T \mathbf{R}^{-1} \nabla_{\mathbf{x}} \mathbf{h} + \boldsymbol{\beta}^T (\mathbf{P}_{k|k-1}^{x})^{-1}.$$

Hence, $\mathcal{F}$ is given by

$$\mathcal{F} = \begin{bmatrix} \nabla_u^T \mathbf{h} \mathbf{R}^{-1} \nabla_u \mathbf{h} & \nabla_u^T \mathbf{h} \mathbf{R}^{-1} \nabla_x \mathbf{h} \\ \nabla_x^T \mathbf{h} \mathbf{R}^{-1} \nabla_u \mathbf{h} & \nabla_x^T \mathbf{h} \mathbf{R}^{-1} \nabla_x \mathbf{h} + (\mathbf{P}_{k|k-1}^x)^{-1} \end{bmatrix}. \tag{14}$$

Putting together the above results yields a nonlinear SISF algorithm named SISF-wDF, which is formally described in Table 1. This algorithm is based on a novel Bayesian perspective of addressing the SISE problem, while the literature usually considers the problem from the viewpoint of filter design and optimal gain selection. It should be pointed out that the SISF-wDF algorithm can be applied to nonlinear systems of general form instead of being restricted to systems of some required special forms.

Remark 1 *(Improvements to the Gauss-Newton method).* While the basic Gauss-Newton iteration shown in (11) solves linear problems within only a single iteration and has fast local convergence on mildly nonlinear problems, it may suffer from divergence for some nonlinear problems. To improve the convergence performance, a damping coefficient $\alpha^{(i)} > 0$ can be added:

$$\hat{\boldsymbol{\xi}}_k^{(i+1)} = \hat{\boldsymbol{\xi}}_k^{(i)} - \alpha^{(i)} [\nabla_\xi^T \mathbf{r}(\hat{\boldsymbol{\xi}}_k^{(i)}) \cdot \nabla_\xi \mathbf{r}(\hat{\boldsymbol{\xi}}_k^{(i)})]^{-1} \cdot \nabla_\xi^T \mathbf{r}(\hat{\boldsymbol{\xi}}_k^{(i)}) \cdot \mathbf{r}(\hat{\boldsymbol{\xi}}_k^{(i)}). \tag{15}$$

One can show that the damped Gauss-Newton iteration keeps moving to the critical point in a descent direction for sufficiently small $\alpha^{(i)} > 0$, thus guaranteeing its local convergence. Yet, $\alpha^{(i)}$ must be selected with caution to ensure the viability

Table 1. The SISF-wDF algorithm (SISF for systems with direct feedthrough).

initialize: $k = 0$, $\hat{\boldsymbol{\xi}}_{0|0} = \mathrm{E}(\boldsymbol{\xi}_0)$, $\mathrm{P}_{0|0}^\xi = p_0 \mathbf{I}$, where p_0 is typically a large positive value
repeat
$k \leftarrow k + 1$
Prediction:
predict the state *via* (6)
compute prediction error covariance *via* (7)
Update:
initialize: $i = 0$, $\hat{\boldsymbol{\xi}}_k^{(0)} = [\mathbf{0}^T\ \hat{\mathbf{x}}_{k|k-1}^T]^T$
while $i < i_{max}$ **do**
perform Gauss-Newton based joint input and state filtering *via* (11)
$i \leftarrow i + 1$
end while
export $\hat{\boldsymbol{\xi}}_k^{(i_{max})}$ and assign to $\hat{\mathbf{u}}_{k|k}$ and $\hat{\mathbf{x}}_{k|k}$
compute joint filtering error covariance *via* (12)–(14)
until no more measurements arrive

of the damped Gauss-Newton, and a few methods have been proposed, e.g., the Armijo-Goldstein step length principle. A further improvement is to introduce a stabilizing term:

$$\hat{\xi}_k^{(i+1)} = \hat{\xi}_k^{(i)} - \alpha^{(i)} [\nabla_\xi^T \mathbf{r}(\hat{\xi}_k^{(i)}) \cdot \nabla_\xi \mathbf{r}(\hat{\xi}_k^{(i)}) + \delta^{(i)} \mathbf{D}^{(i)}]^{-1} \cdot \nabla_\xi^T \mathbf{r}(\hat{\xi}_k^{(i)}) \cdot \mathbf{r}(\hat{\xi}_k^{(i)}),$$

whereby the rank deficiency problem of $\nabla_\xi^T \mathbf{r} \nabla_\xi \mathbf{r}$ that may appear in (11) and (15) can be avoided, given that $\delta^{(i)} > 0$ and $\mathbf{D}^{(i)}$ is a specified symmetric positive definite matrix. This is known as the trust region method or Levenberg-Marquardt method. For more details about Gauss-Newton-type methods, the reader is referred to (Björck 1996).

Remark 2 *(Generality of the Bayesian SISF algorithm).* The work (Fang et al. 2013) proves that the SISF-wDF algorithm, if applied to a linear system with direct feedthrough, will give the same input and state estimation as in (Gillijns and De Moor 2007b). The method in (Gillijns and De Moor 2007b) is based on MVUE and the development of the SISF-wDF algorithm provides its statistical interpretation. That is, it is statistically optimal if the following assumptions hold: (1) $\mathbf{x}_0 \sim \mathcal{N}$ $(\hat{\mathbf{x}}_{0|0}, \mathbf{P}_{0|0}^x)$; (2) $\{\mathbf{w}_k\}$ and $\{\mathbf{v}_k\}$ are zero-mean white Gaussian; (3) $\mathbf{x}_0$, $\{\mathbf{w}_k\}$ and $\{\mathbf{v}_k\}$ are independent of each other; (4) $\{\mathbf{u}_k\}$ is white Gaussian and independent of $\mathbf{x}_0$, $\{\mathbf{w}_k\}$ and $\{\mathbf{v}_k\}$. Compared to the assumptions made for the classical KF, (4) is the only additional one, which ensures that the state propagation and output measurement sequences are Gaussian distributed. It is noteworthy that, even though the derivation of the method in (Gillijns and De Moor 2007b) proceeds without imposing the assumption (A1), its statistical optimality still implicitly relies on this assumption. We note that the observations here partially reflect the fact that Bayesian estimation can offer a general framework to solve SISE problems.

3. SISF for Systems without Direct Feedthrough

In this section, we extend the results of Section 2 and consider a nonlinear system described by equations of the following form:

$$\begin{cases} \mathbf{x}_{k+1} = \mathbf{f}(\mathbf{x}_k, \mathbf{u}_k) + \mathbf{w}_k, \\ \mathbf{y}_k = \mathbf{h}(\mathbf{x}_k) + \mathbf{v}_k, \end{cases} \tag{16}$$

where no direct input-to-output feedthrough exists. In this situation, the input estimation has to be delayed by one time step because the first measurement containing information about $\mathbf{u}_{k-1}$ is $\mathbf{y}_k$. Therefore, it is $p(\mathbf{u}_{k-1}, \mathbf{x}_k | \mathbf{Y}_{1:k})$ that is of interest here and should be sequentially updated. We impose the same assumption as (A1) to $\mathbf{u}_k$ for the system in (16), i.e., $\{\mathbf{u}_k\}$ is a white process independent of $\mathbf{x}_0$, $\{\mathbf{w}_k\}$ and $\{\mathbf{v}_k\}$. Using the Bayes' rule, we can construct a Bayesian SISF paradigm for this case. Omitting the intermediate steps,

$$p(\mathbf{u}_{k-1}, \mathbf{x}_k|\mathbf{Y}_{1:k}) \propto p(\mathbf{y}_k|\mathbf{x}_k)\int p(\mathbf{x}_k|\mathbf{u}_{k-1}, \mathbf{x}_{k-1}) \cdot p(\mathbf{x}_{k-1}|\mathbf{Y}_{1:k-1})d\mathbf{x}_{k-1}. \tag{17}$$

We also introduce the following assumptions in order to formulate a tractable MAP-based SISF problem:

$$p(\mathbf{u}_{k-1}, \mathbf{x}_k|\mathbf{Y}_k) \sim \mathcal{N}\left(\begin{bmatrix} \hat{\mathbf{u}}_{k-1|k} \\ \hat{\mathbf{x}}_{k|k} \end{bmatrix}, \begin{bmatrix} \mathbf{P}^{\mathbf{u}}_{k-1|k} & \mathbf{P}^{\mathbf{ux}}_{k-1,k|k} \\ (\mathbf{P}^{\mathbf{ux}}_{k-1,k|k})^T & \mathbf{P}^{\mathbf{x}}_{k|k} \end{bmatrix}\right),$$

$$p(\mathbf{y}_k|\mathbf{x}_k) \sim \mathcal{N}(\mathbf{h}(\mathbf{x}_k), \mathbf{R}_k).$$

From (17), a MAP cost function can be defined as done previously in (9), which is

$$\ell(\mathbf{u}_{k-1}, \mathbf{x}_k) = \delta + \mathbf{r}^T(\mathbf{u}_{k-1}, \mathbf{x}_k) \cdot \mathbf{r}(\mathbf{u}_{k-1}, \mathbf{x}_k). \tag{18}$$

Here, δ is a constant and

$$\mathbf{r}(\mathbf{u}_{k-1}, \mathbf{x}_k) = \begin{bmatrix} \mathbf{R}_k^{-\frac{1}{2}} \boldsymbol{\rho}_k \\ \boldsymbol{\Pi}^{-\frac{1}{2}}(\mathbf{u}_{k-1})\boldsymbol{\zeta}_k \end{bmatrix},$$

where $\boldsymbol{\rho}_k = \mathbf{y}_k - \mathbf{h}(\mathbf{x}_k)$, $\boldsymbol{\zeta}_k = \mathbf{x}_k - \mathbf{f}(\mathbf{u}_{k-1}, \hat{\mathbf{x}}_{k-1|k-1})$, and $\boldsymbol{\Pi}(\mathbf{u}_{k-1}) = \nabla_{\mathbf{x}}\mathbf{f}(\mathbf{u}_{k-1}, \hat{\mathbf{x}}_{k-1|k-1}) \cdot \mathbf{P}^{\mathbf{x}}_{k-1|k-1} \cdot \nabla_{\mathbf{x}}^T\mathbf{f}(\mathbf{u}_{k-1}, \hat{\mathbf{x}}_{k-1|k-1}) + \mathbf{Q}_{k-1}$. For this problem, a numerical solution can also be developed using the Gauss-Newton method through the following iterative procedure:

$$\hat{\boldsymbol{\sigma}}_k^{(i+1)} = \hat{\boldsymbol{\sigma}}_k^{(i)} - [\nabla_{\boldsymbol{\sigma}}^T\mathbf{r}(\hat{\boldsymbol{\sigma}}_k^{(i)}) \cdot \nabla_{\boldsymbol{\sigma}}\mathbf{r}(\hat{\boldsymbol{\sigma}}_k^{(i)})]^{-1} \cdot \nabla_{\boldsymbol{\sigma}}^T\mathbf{r}(\hat{\boldsymbol{\sigma}}_k^{(i)}) \cdot \mathbf{r}(\hat{\boldsymbol{\sigma}}_k^{(i)}), \tag{19}$$

where $\boldsymbol{\sigma}_k = [\mathbf{u}_{k-1}^T \ \mathbf{x}_k^T]^T$ and $\nabla_{\boldsymbol{\sigma}}\mathbf{r} = [\nabla_{\mathbf{u}}\mathbf{r} \ \nabla_{\mathbf{x}}\mathbf{r}]$. Here, one can set the initial condition as $\hat{\boldsymbol{\sigma}}_k^{(0)} = \mathbf{0}$, run the procedure iteratively, and finally assign the obtained $\hat{\boldsymbol{\sigma}}_k^{(i_{max})}$ to $\hat{\mathbf{u}}_{k-1|k}$ and $\hat{\mathbf{x}}_{k|k}$.

Proceeding further, the associated estimation error covariance matrix can be computed by evaluating the Fisher information matrix at $\hat{\mathbf{u}}_{k-1|k}$ and $\hat{\mathbf{x}}_{k|k}$, i.e.,

$$\begin{bmatrix} \mathbf{P}^{\mathbf{u}}_{k-1|k} & \mathbf{P}^{\mathbf{ux}}_{k-1,k|k} \\ (\mathbf{P}^{\mathbf{ux}}_{k-1,k|k})^T & \mathbf{P}^{\mathbf{x}}_{k|k} \end{bmatrix} = \mathcal{F}^{-1}(\hat{\mathbf{u}}_{k-1|k}, \hat{\mathbf{x}}_{k|k}), \tag{20}$$

where the definition of $\mathcal{F}$ is identical to (13). Fully determining each block of $\mathcal{F}$ entails the computation as below.

The *l*-th column of the gradient matrix of $\mathbf{r}$ with respect to (w.r.t.) $\mathbf{u}$ is given by,

$$\frac{\partial \mathbf{r}}{\partial \mathbf{u}_l} = \begin{bmatrix} \mathbf{0} \\ -\boldsymbol{\Pi}^{-\frac{1}{2}}\frac{\partial \mathbf{f}}{\partial \mathbf{u}_l} - \frac{1}{2}\boldsymbol{\Pi}^{\frac{1}{2}}\boldsymbol{\Pi}^{-1}\frac{\partial \boldsymbol{\Pi}}{\partial \mathbf{u}_l}\boldsymbol{\Pi}^{-1}\boldsymbol{\zeta} \end{bmatrix}.$$

The following relation is used here:

$$\frac{\partial \mathbf{x}^{\frac{1}{2}}}{\partial \tau} = -\frac{1}{2}\mathbf{X}^{\frac{1}{2}}\mathbf{X}^{-1}\frac{\partial \mathbf{x}}{\partial \tau}\mathbf{X}^{-1},$$

where $\mathbf{X}$ is a symmetric positive definite matrix dependent on τ (Spinello and Stilwell 2010). The l-th column of the gradient matrix of $\mathbf{r}$ w.r.t. $\mathbf{x}$ is

$$\frac{\partial \mathbf{r}}{\partial \mathbf{x}_l} = \begin{bmatrix} -\mathbf{R}^{-\frac{1}{2}}\dfrac{\partial \mathbf{h}}{\partial \mathbf{x}_l} \\ \mathbf{\Pi}^{-\frac{1}{2}}\mathbf{e}_l \end{bmatrix}, \nabla_{\mathbf{x}}\,\mathbf{r} = \begin{bmatrix} -\mathbf{R}^{-\frac{1}{2}}\nabla_{\mathbf{x}}\mathbf{h} \\ \mathbf{\Pi}^{-\frac{1}{2}} \end{bmatrix},$$

where $\mathbf{e}_l$ is the standard basis vector with a 1 in the l-th element and 0's elsewhere. The lj-th entries of $\nabla_{\mathbf{u}}^T\,\mathbf{r}\nabla_{\mathbf{u}}\mathbf{r}$, $\nabla_{\mathbf{u}}^T\,\mathbf{r}\nabla_{\mathbf{x}}\mathbf{r}$ and $\nabla_{\mathbf{x}}^T\,\mathbf{r}\nabla_{\mathbf{x}}\mathbf{r}$ are expressed as, respectively,

$$\frac{\partial \mathbf{r}^T}{\partial \mathbf{u}_l}\frac{\partial \mathbf{r}}{\partial \mathbf{u}_j} =$$

$$\frac{\partial \mathbf{f}^T}{\partial \mathbf{u}_l}\mathbf{\Pi}^{-1}\frac{\partial \mathbf{f}}{\partial \mathbf{u}_j} + \frac{1}{2}\zeta^T\mathbf{\Pi}^{-1}\cdot\left(\frac{\partial \mathbf{\Pi}}{\partial \mathbf{u}_l}\mathbf{\Pi}^{-1}\frac{\partial \mathbf{f}}{\partial \mathbf{u}_j} + \frac{\partial \mathbf{\Pi}}{\partial \mathbf{u}_j}\mathbf{\Pi}^{-1}\frac{\partial \mathbf{f}}{\partial \mathbf{u}_l}\right) + \frac{1}{4}\zeta^T\mathbf{\Pi}^{-1}\frac{\partial \mathbf{\Pi}}{\partial \mathbf{u}_l}\mathbf{\Pi}^{-1}\frac{\partial \mathbf{\Pi}}{\partial \mathbf{u}_j}\mathbf{\Pi}^{-1}\zeta,$$

$$\frac{\partial \mathbf{r}^T}{\partial \mathbf{u}_l}\frac{\partial \mathbf{r}}{\partial \mathbf{x}_j} = -\frac{\partial \mathbf{f}^T}{\partial \mathbf{u}_l}\mathbf{\Pi}^{-1}\mathbf{e}_j - \frac{1}{2}\zeta^T\mathbf{\Pi}^{-1}\frac{\partial \mathbf{\Pi}}{\partial \mathbf{u}_l}\mathbf{\Pi}^{-1}\mathbf{e}_j,$$

$$\frac{\partial \mathbf{r}^T}{\partial \mathbf{x}_l}\frac{\partial \mathbf{r}}{\partial \mathbf{x}_j} = \frac{\partial \mathbf{h}^T}{\partial \mathbf{x}_l}\mathbf{R}^{-1}\frac{\partial \mathbf{h}}{\partial \mathbf{x}_j} + \mathbf{e}_l^T\mathbf{\Pi}^{-1}\mathbf{e}_j,$$

$$\nabla_{\mathbf{x}}^T\mathbf{r}\nabla_{\mathbf{x}}\mathbf{r} = \nabla_{\mathbf{x}}^T\mathbf{h}\mathbf{R}^{-1}\nabla_{\mathbf{x}}\mathbf{h} + \mathbf{\Pi}^{-1}.$$

Then, we have

$$\frac{\partial \ell}{\partial \mathbf{u}_l} = \mathbf{r}^T\frac{\partial \mathbf{r}}{\partial \mathbf{u}_l} = -\zeta^T\mathbf{\Pi}^{-1}\frac{\partial \mathbf{f}}{\partial \mathbf{u}_l} - \frac{1}{2}\zeta^T\mathbf{\Pi}^{-1}\frac{\partial \mathbf{\Pi}}{\partial \mathbf{u}_l}\mathbf{\Pi}^{-1}\zeta,$$

$$\frac{\partial \ell}{\partial \mathbf{x}_l} = \mathbf{r}^T\frac{\partial \mathbf{r}}{\partial \mathbf{x}_l} = -\rho^T\mathbf{R}^{-1}\frac{\partial \mathbf{h}}{\partial \mathbf{x}_l} + \zeta^T\mathbf{\Pi}^{-1}\mathbf{e}_l,$$

$$\nabla_{\mathbf{x}}\ell = -\rho^T\mathbf{R}^{-1}\nabla_{\mathbf{x}}\mathbf{h} + \zeta^T\mathbf{\Pi}^{-1}.$$

To compute the Fisher information matrix, $E(\nabla_{\mathbf{u}}^T\ell\nabla_{\mathbf{u}}\ell)$, $E(\nabla_{\mathbf{u}}^T\ell\nabla_{\mathbf{x}}\ell)$, and $E(\nabla_{\mathbf{x}}^T\ell\nabla_{\mathbf{x}}\ell)$ are needed. Their lj-th entries are

$$E\left(\frac{\partial \ell^T}{\partial \mathbf{u}_l}\frac{\partial \ell}{\partial \mathbf{u}_j}\right) = \frac{\partial \mathbf{f}^T}{\partial \mathbf{u}_l}\mathbf{\Pi}^{-1}\frac{\partial \mathbf{f}}{\partial \mathbf{u}_j} + \frac{1}{4}\operatorname{tr}\left(\frac{\partial \mathbf{\Pi}}{\partial \mathbf{u}_l}\mathbf{\Pi}^{-1}\frac{\partial \mathbf{\Pi}}{\partial \mathbf{u}_j}\mathbf{\Pi}^{-1}\right),$$

$$E\left(\frac{\partial \ell^T}{\partial \mathbf{u}_l}\frac{\partial \ell}{\partial \mathbf{x}_j}\right) = -\frac{\partial \mathbf{f}^T}{\partial \mathbf{u}_l}\mathbf{\Pi}^{-1}\mathbf{e}_j,$$

$$E\left(\frac{\partial \ell^T}{\partial \mathbf{x}_l}\frac{\partial \ell}{\partial \mathbf{x}_j}\right) = \frac{\partial \mathbf{h}^T}{\partial \mathbf{x}_l}\mathbf{\Pi}^{-1}\frac{\partial \mathbf{h}}{\partial \mathbf{x}_j} + \mathbf{e}_l^T\mathbf{\Pi}^{-1}\mathbf{e}_j,$$

$$E\left(\nabla_{\mathbf{x}}^T\ell\nabla_{\mathbf{x}}\ell\right) = \nabla_{\mathbf{x}}^T\mathbf{h}\mathbf{R}^{-1}\nabla_{\mathbf{x}}\mathbf{h} + \mathbf{\Pi}^{-1}.$$

With the above derivation, we have fully developed the SISF algorithm for the system in (16), which is named `SISF-w/oDF` and summarized in Table 2. Note that, if applied to a linear system, the `SISF-w/oDF` algorithm will reduce to (Gillijns and De Moor 2007a), which offers a MVUE-based SISF algorithm for linear systems without direct feedthrough. In other words, if the conditions (1)–(4) proposed in Remark 2 are also valid for the linear version of (16), the algorithm in (Gillijns and De Moor 2007a) can be directly derived using the Bayesian paradigm along with MAP estimation. In addition (Gillijns and De Moor 2007a) gives the same state update as (Kitanidis 1987, Darouach and Zasadzinski 1997) and the same input update as (Hsieh 2000). This suggests that these methods can be regarded as special cases of the `SISF-w/oDF` algorithm.

Table 2. The `SISF-w/oDF` algorithm (SISF for systems without direct feedthrough).

initialize: $k = 0$, $\hat{\sigma}_0 = E(\sigma_0)$, $\mathbf{P}_0^\sigma = p_0\mathbf{I}$, where p_0 is a large positive value
repeat
$k \leftarrow k + 1$
initialize: $i = 0$, $\hat{\sigma}_k^{(0)} = \mathbf{0}$
while $i < i_{max}$ **do**
perform Gauss-Newton based joint input and state filtering *via* (19)
$i \leftarrow i + 1$
end while
export $\hat{\sigma}_k^{(i_{max})}$ and assign to $\hat{\mathbf{u}}_{k-1|k}$ and $\hat{\mathbf{x}}_{k|k}$
compute joint filtering error covariance *via* (20)
until no more measurements arrive

4. SISS for Systems with and without Direct Feedthrough

Our results above are concerned with the filtering problem, where $\mathbf{u}_k$ (or $\mathbf{u}_{k-1}$) and $\mathbf{x}_k$ are estimated based on the measurement $\mathbf{Y}_{1:k}$. Another interesting problem is to make the estimation when all the measurements are available. Suppose that the total time is N. Then, the question is to make an estimate of $\mathbf{u}_k$ (or $\mathbf{u}_{k-1}$) and $\mathbf{x}_k$ using $\mathbf{Y}_{1:N}$. This is a problem of smoothing or SISS as termed before, which is the theme of this section. Following a similar structure as Sections 2 and 3, we will first investigate Bayesian SISS for systems with direct feedthrough and then move forward to those without it. Here, we would like to point out that developing effective and efficient enough SISS algorithms for nonlinear systems without direct feedthrough is rather difficult because of the asynchronous coupling between $\mathbf{u}_{k-1}$ and $\mathbf{x}_k$ and the nonlinearity involved. We, thus, focus on only linear systems in this case.

4.1 SISS for Nonlinear Systems with Direct Feedthrough

Consider the nonlinear system in (1). Section 2 discusses SISF for (1) from the viewpoint of determining $p(\mathbf{u}_k, \mathbf{x}_k|\mathbf{Y}_{1:k})$, or more specifically, the passing from $p(\mathbf{u}_{k-1}, \mathbf{x}_{k-1}|\mathbf{Y}_{1:k-1})$ to $p(\mathbf{u}_k, \mathbf{x}_k|\mathbf{Y}_{1:k})$. Here, we shift our attention to $p(\mathbf{u}_k, \mathbf{x}_k|\mathbf{Y}_{1:N})$ in order to achieve fixed-interval SISS. Because the reader has been familiar with the notation, we use $\xi_k = [\mathbf{u}_k^T \ \mathbf{x}_k^T]^T$ to improve notational simplicity. Applying the Bayes' rule, we can build the Bayesian SISS paradigm for (1), which unveils the backward recursion of $p(\xi_k|\mathbf{Y}_{1:N})$ from $p(\mathbf{x}_{k+1}|\mathbf{Y}_{1:N})$:

$$p(\xi_k|\mathbf{Y}_{1:N}) = p(\xi_k|\mathbf{Y}_{1:k})\int \frac{p(\mathbf{x}_{k+1}|\xi_k)\cdot p(\mathbf{x}_{k+1}|\mathbf{Y}_{1:N})}{p(\mathbf{x}_{k+1}|\mathbf{Y}_{1:k})} \ d\mathbf{x}_{k+1}. \tag{21}$$

Here, we demonstrate the derivation. First,

$$p(\xi_k|\mathbf{Y}_{1:N}) = \int p(\xi_k, \xi_{k+1}|\mathbf{Y}_{1:N})d\xi_{k+1} = \int p(\xi_k|\xi_{k+1}, \mathbf{Y}_{1:N}) \cdot p(\xi_{k+1}|\mathbf{Y}_{1:N})d\xi_{k+1}.$$

Note that ξ_k is conditionally independent of $\mathbf{Y}_{k+1:N}$ and $\mathbf{u}_{k+1}$ given $\mathbf{x}_{k+1}$, due to the Markovian state propagation, and (A1). Hence, $p(\xi_k|\xi_{k+1}, \mathbf{Y}_{1:N}) = p(\xi_k|\mathbf{x}_{k+1}, \mathbf{Y}_{1:k})$. Then, we have

$$p(\xi_k|\mathbf{Y}_{1:N}) = \int p(\xi_k|\mathbf{x}_{k+1}, \mathbf{Y}_{1:k}) \cdot p(\xi_{k+1}|\mathbf{Y}_{1:N})d\xi_{k+1}$$
$$= \int p(\xi_k|\mathbf{x}_{k+1}, \mathbf{Y}_{1:k}) \cdot p(\mathbf{x}_{k+1}|\mathbf{Y}_{1:N})d\mathbf{x}_{k+1}. \tag{22}$$

Meanwhile, we have

$$p(\xi_k|\mathbf{x}_{k+1},\mathbf{Y}_{1:k}) = \frac{p(\xi_k, \mathbf{x}_{k+1}, \mathbf{Y}_{1:k})}{p(\mathbf{x}_{k+1}, \mathbf{Y}_{1:k})} = \frac{p(\mathbf{x}_{k+1}|\xi_k, \mathbf{Y}_{1:k}) \cdot p(\xi_k|\mathbf{Y}_{1:k})}{p(\mathbf{x}_{k+1}|\mathbf{Y}_{1:k})} \tag{23}$$

$$= \frac{p(\mathbf{x}_{k+1}|\xi_k) \cdot p(\xi_k|\mathbf{Y}_{1:k})}{p(\mathbf{x}_{k+1}|\mathbf{Y}_{1:k})}.$$

Inserting (23) into (22), we can obtain (21).

The Bayesian paradigm in (21) is an input and state smoother in a statistical sense, which illustrates the backward update of ξ_k given $\mathbf{Y}_{1:N}$. However, a direct analytical evaluation of the pdf's is known to be quite difficult, if not impossible, for nonlinear systems. Hence, we will formulate an MAP estimation problem based on (21) and seek a numerical solution. To proceed further, we make Gaussian distribution assumptions for the following pdf's:

(A5) $p(\mathbf{x}_{k+1}|\xi_k) \sim \mathcal{N}(\mathbf{f}(\xi_k), \mathbf{Q}_k)$;

(A6) $p(\xi_k|\mathbf{Y}_{1:k}) \sim \mathcal{N}(\hat{\xi}_{k|k}, \mathbf{P}^\xi_{k|k})$;

(A7) $p(\mathbf{x}_{k+1}|\mathbf{Y}_{1:N}) \sim \mathcal{N}(\hat{\mathbf{x}}_{k+1|N}, \mathbf{P}^\mathbf{x}_{k+1|N})$;

(A8) $p(\mathbf{x}_{k+1}|\mathbf{Y}_{1:k}) \sim \mathcal{N}(\hat{\mathbf{x}}_{k+1|k}, \mathbf{P}^\mathbf{x}_{k+1|k})$.

Here, $\hat{\xi}_{k|k}$ is the filtered estimate of ξ_k given $\mathbf{Y}_{1:k}$, $\mathbf{P}^\xi_{k|k}$ is the filtering error covariance, $\hat{\mathbf{x}}_{k+1|N}$ is the smoothed estimate of $\mathbf{x}_{k+1}$ given $\mathbf{Y}_{1:N}$, and $\mathbf{P}^\mathbf{x}_{k+1|N}$ is the smoothing error covariance. These assumptions are made to bridge the gap from the Bayesian paradigm in (21) to an executable smoother.

We now consider developing an MAP-based smoother to estimate $\mathbf{u}_k$ and $\mathbf{x}_k$ via maximizing $p(\xi_k|\mathbf{Y}_{1:N})$. The smoother then can be expressed as:

$$\hat{\xi}_{k|N} = \arg\max_{\xi_k} p(\xi_k|\mathbf{Y}_{1:N}). \tag{24}$$

The above maximization of $p(\xi_k|\mathbf{Y}_{1:N})$ can be transformed into the following problem of minimizing a cost function with assistance of the assumptions (A5)–(A8), which is given as:

$$\hat{\xi}_{k|N} = \arg\min_{\xi_k} \ell(\xi_k), \tag{25}$$

where

$$\ell(\xi_k) := (\xi_k - \hat{\xi}_{k|k})^T (\mathbf{P}^\xi_{k|k})^{-1} (\xi_k - \hat{\xi}_{k|k}) + (\mathbf{f}(\xi_k) - \delta_k)^T \Delta_k^{-1} (\mathbf{f}(\xi_k) - \delta_k), \tag{26}$$

$$\Delta_k = [(\mathbf{P}^\mathbf{x}_{k+1|N})^{-1} - (\mathbf{P}^\mathbf{x}_{k+1|k})^{-1}]^{-1} + \mathbf{Q}_k, \tag{27}$$

$$\delta_k = [(\mathbf{P}^\mathbf{x}_{k+1|N})^{-1} - (\mathbf{P}^\mathbf{x}_{k+1|k})^{-1}]^{-1} \cdot [(\mathbf{P}^\mathbf{x}_{k+1|N})^{-1} \hat{\mathbf{x}}_{k+1|N} - (\mathbf{P}^\mathbf{x}_{k+1|k})^{-1} \hat{\mathbf{x}}_{k+1|k}]. \tag{28}$$

For a detailed derivation of the above, an interested reader is referred to (Fang et al. 2015). Furthermore, the sum of the weighted 2-norms in (4.1) can be rewritten as

$$\ell(\xi_k) = \mathbf{r}^T(\xi_k) \cdot \mathbf{r}(\xi_k), \tag{29}$$

where

$$\mathbf{r}(\xi_k) = \begin{bmatrix} (\mathbf{P}_{k|k}^{\xi})^{-\frac{1}{2}}(\xi_k - \hat{\xi}_{k|k}) \\ \Delta_k^{-\frac{1}{2}}(\mathbf{f}(\xi_k) - \delta_k) \end{bmatrix}.$$

This would allow the use of the classical Gauss-Newton method, which performs an iterative searching process that linearizes around the current arrival point, to determine the best search direction and then to move forward to the next point. Specifically, we have

$$\hat{\xi}_{k|N}^{(i+1)} = \hat{\xi}_{k|N}^{(i)} - [\nabla_\xi^T \mathbf{r}(\hat{\xi}_{k|N}^{(i)}) \cdot \nabla_\xi \mathbf{r}(\hat{\xi}_{k|N}^{(i)})]^{-1} \cdot \nabla_\xi^T \mathbf{r}(\hat{\xi}_{k|N}^{(i)}) \cdot \mathbf{r}(\hat{\xi}_{k|N}^{(i)}), \tag{30}$$

where (i) denotes the iteration number, and

$$\nabla_\xi \mathbf{r}(\xi_k) = \begin{bmatrix} (\mathbf{P}_{k|k}^{\xi})^{-\frac{1}{2}} \\ \Delta_k^{-\frac{1}{2}}\nabla_\xi \mathbf{f}(\xi_k) \end{bmatrix}.$$

One can let $\hat{\xi}_{k|N} = \hat{\xi}_{k|N}^{(i\max)}$, where $i_{\max}$ is the maximum number of iterations. The Fisher information matrix $\mathcal{F}$ for (4.2) is approximately given by

$$\mathcal{F}(\xi_k) = (\mathbf{P}_{k|k}^{\xi})^{-1} + \nabla_\xi^T \mathbf{f}(\xi_k)\Delta_k^{-1}\nabla_\xi \mathbf{f}(\xi_k).$$

Evaluating $\mathcal{F}$ at $\hat{\xi}_{k|N}$ and inverting it will lead to the error covariance, i.e.,

$$\mathbf{P}_{k|N}^{\xi} = \mathcal{F}^{-1}(\hat{\xi}_{k|N}) = [(\mathbf{P}_{k|k}^{\xi})^{-1} + \nabla_\xi^T \mathbf{f}(\hat{\xi}_{k|N})\Delta_k^{-1}\nabla_\xi \mathbf{f}(\hat{\xi}_{k|N})]^{-1}. \tag{31}$$

We say that (30)–(31) are the backward smoothing equations for input and state estimation. The corresponding forward filtering equations are given by the `SISF-wDF` algorithm. The above nonlinear SISS algorithm can be readily specialized to a linear case. Consider a linear system of the following form

$$\begin{cases} \mathbf{x}_{k+1} = \mathbf{F}_k\mathbf{x}_k + \mathbf{G}_k\mathbf{u}_k + \mathbf{w}_k, \\ \mathbf{y}_k = \mathbf{H}_k\mathbf{x}_k + \mathbf{J}_k\mathbf{u}_k + \mathbf{v}_k. \end{cases} \tag{32}$$

Then, the smoother in (30)–(31) can find the best input and state estimates in a single iteration, which leads to a linear SISS algorithm. It can be proven that the algorithm can be expressed as

$$\hat{\xi}_{k|N} = \hat{\xi}_{k|k} + \mathbf{K}_k(\hat{\mathbf{x}}_{k+1|N} - \bar{\mathbf{F}}_k\hat{\xi}_{k|k}), \tag{33}$$

$$\mathbf{K}_k = \mathbf{P}^\xi_{k|k}\,\bar{\mathbf{F}}_k^T\,(\mathbf{P}^\mathbf{x}_{k+1|k})^{-1}, \tag{34}$$

where, $\bar{\mathbf{F}}_k = [\mathbf{G}_k\ \mathbf{F}_k]$. The associated smoothing error covariance is given by

$$\mathbf{P}^\xi_{k|N} = \mathbf{P}^\xi_{k|k} + \mathbf{K}_k(\mathbf{P}^\mathbf{x}_{k+1|N} - \mathbf{P}^\mathbf{x}_{k+1|k})\,\mathbf{K}_k^T. \tag{35}$$

To sum up, we call the obtained forward-backward smoothing algorithm as `SISS-wDF` and summarize it in Table 3.

4.2 SISS for Systems without Direct Feedthrough

Now, let us consider Bayesian SISS for systems without direct feedthrough, which are shown in (16). In this case, we use the notation $\sigma_k = [\mathbf{u}_{k-1}^T\ \mathbf{x}_k^T]^T$. Given the assumption (A1), the Bayesian smoothing paradigm for input and state estimation is given by (Fang et al. 2015):

$$\mathbf{p}(\sigma_k|\mathbf{Y}_{1:N}) = p(\sigma_k|\mathbf{Y}_{1:k})\iint \frac{p(\mathbf{x}_{k+1}|\mathbf{x}_k, \mathbf{u}_k)\cdot p(\sigma_{k+1}|\mathbf{Y}_{1:N})}{\int p(\mathbf{x}_{k+1}|\mathbf{x}_k, \mathbf{u}_k)\cdot p(\mathbf{x}_k|\mathbf{Y}_{1:k})d\mathbf{x}_k}\,d\mathbf{u}_k d\mathbf{x}_{k+1}. \tag{36}$$

Table 3. The `SISS-wDF` algorithm: SISS for systems with direct feedthrough.

Forward filtering:
for $k = 1$ to N **do**
compute filtered input and state estimate *via* the `SISF-wDF` algorithm
end for
Backward smoothing:
for $k = N-1$ to 1 **do**
initialize: $i = 0$, $\hat{\xi}^{(0)}_{k|N} = \hat{\xi}_{k|k}$
while $i < i_{\max}$ **do**
perform Gauss-Newton-based joint input and state smoothing *via* (30) (or (33)–(34) for linear systems)
$i \leftarrow i + 1$
end while
export $\hat{\mathbf{u}}_{k|N}$ and $\hat{\mathbf{x}}_{k|N}$ from $\hat{\xi}^{(i_{\max})}_{k|N}$
compute joint smoothing error covariance *via* (31) (or (35) for linear systems)
end for

Our research shows that it is challenging to derive a viable SISS method based on (36) for a nonlinear system of general form as shown in (16). This is because $\mathbf{u}_{k-1}$ and $\mathbf{x}_k$ are asynchronous in time with a gap of one time step between them. The asynchronous coupling complicates the nonlinear relationship between $\mathbf{u}_{k-1}$, $\mathbf{x}_k$, and the output measurements, which is only exacerbated further in the backward smoothing scenario. Hence, we instead constrain our scope to linear systems. Consider

$$\begin{cases} \mathbf{x}_{k+1} = \mathbf{F}_k\mathbf{x}_k + \mathbf{G}_k\mathbf{u}_k + \mathbf{w}_k, \\ \mathbf{y}_k = \mathbf{H}_k\mathbf{x}_k + \mathbf{v}_k. \end{cases} \tag{37}$$

For (37), the MAP smoother $\hat{\sigma}_{k|N} = \text{argmax}_{\sigma_k}\, p(\sigma_k|\mathbf{Y}_{1:N})$ can be converted into the following problem:

$$\hat{\sigma}_{k|N} = \underset{\sigma_k}{\text{argmin}}\ \ell(\sigma_k), \tag{38}$$

where

$$\ell(\sigma_k) := (\sigma_k - \hat{\sigma}_{k|k})^T (\mathbf{P}_{k|k}^\sigma)^{-1} (\sigma_k - \hat{\sigma}_{k|k}) + (\mathbf{F}_k\mathbf{N}\sigma_k - \delta_k)^T \Delta_k^{-1} (\mathbf{F}_k\mathbf{N}\sigma_k - \delta_k), \tag{39}$$

$$\mathbf{M}_k = [\ \mathbf{G}_k\ \mathbf{I}], \qquad \mathbf{N} = [\mathbf{0}\ \mathbf{I}], \tag{40}$$

$$\Delta_k = [(\mathbf{M}_k\,\mathbf{P}_{k+1|N}^\sigma\mathbf{M}_k^T)^{-1} - (\mathbf{P}_{k+1|k}^{\mathbf{x}*})^{-1}]^{-1} + \mathbf{Q}_k \tag{41}$$

$$\delta_k = \mathbf{F}_k\,\hat{\mathbf{x}}_{k|k} - \mathbf{P}_{k+1|k}^{\mathbf{x}*}\,(\mathbf{M}_k\,\mathbf{P}_{k+1|N}^\sigma\,\mathbf{M}_k^T - \mathbf{P}_{k+1|k}^{\mathbf{x}*})^{-1}\,(\mathbf{M}_k\,\hat{\sigma}_{k+1|N} - \mathbf{F}_k\,\hat{\mathbf{x}}_{k|k}), \tag{42}$$

$$\mathbf{P}_{k+1|k}^{\mathbf{x}*} = \mathbf{F}_k\,\mathbf{P}_{k|k}^{\mathbf{x}}\,\mathbf{F}_k^T + \mathbf{Q}_k. \tag{43}$$

The solution to (38) can then be derived as follows:

$$\mathbf{K}_k = \mathbf{P}_{k|k}^\sigma\,\mathbf{N}^T\,\mathbf{F}_k^T\,(\mathbf{P}_{k+1|k}^{\mathbf{x}*})^{-1}, \tag{44}$$

$$\hat{\sigma}_{k|N} = \hat{\sigma}_{k|k} + \mathbf{K}_k(\mathbf{M}_k\,\hat{\sigma}_{k+1|N} - \mathbf{F}_k\hat{\mathbf{x}}_{k|k}), \tag{45}$$

$$\mathbf{P}_{k|N}^\sigma = \mathbf{P}_{k|k}^\sigma + \mathbf{K}_k[\mathbf{M}_k\mathbf{P}_{k+1|N}^\sigma\,\mathbf{M}_k^T - (\mathbf{P}_{k+1|k}^{\mathbf{x}*})]^{-1}\,\mathbf{K}_k^T. \tag{46}$$

In above, we formulate and solve an MAP-optimization-based SISS problem for linear systems without direct feedthrough. The backward smoothing solution is offered in (44)–(46) and its associated forward filtering can be accomplished by running the SISF-w/oDF algorithm, which only requires a single iteration. Please note that application of the SISF-w/oDF algorithm to linear systems leads to the joint input and state estimator proposed in (Gillijns and De Moor 2007a), as mentioned in Section 3. Finally, we outline the obtained SISS algorithm, named SISS-w/oDF, in Table 4.

Table 4. The `SISS-w/oDF` algorithm: SISS for linear systems without direct feedthrough.

Forward filtering:
for $k = 1$ to N **do**
compute filtered input and state estimate *via* the `SISF-w/oDF` algorithm or the algorithm in (Gillijns and De Moor, 2007a)
end for
Backward smoothing:
for $k = N - 1$ to 1 **do**
perform joint input and state smoothing *via* (44)–(45)
export $\hat{\mathbf{u}}_{k-1|N}$ and $\hat{\mathbf{x}}_{k|N}$ from $\hat{\sigma}_{k|N}$
compute joint smoothing error covariance *via* (46)
end for

5. Application Example

Ocean flow field reconstruction has been a research subject of intense interest for its vital role in helping oceanographers understand the oceans. Flows are known to be crucial for fishing, shipping, navigation, weather forecasting, environmental monitoring, and climate change. To study these flows, consider a swarm of inexpensive buoyancy-controlled profiling floats acting as an ocean observing system (Colgan 2006, Ouimet and Cortés 2014, Han et al. 2010), deployed to traverse a region of the ocean. The floats are capable of arbitrary vertical migration while traveling along the flows. During the travel, each float measures and stores a time record of its depth, acceleration, position, and some oceanographic quantities such as temperature and salinity. The data record is sent to a central server for processing when the float comes up to the surface. Here, we only consider a three-dimensional flow domain (see Figure 2), the space of which is occupied mainly by two adjacent eddies of opposite direction of rotation.

5.1 Float Dynamics

For simplicity, we only examine a float's motion along the x-direction. The same results can apply to y-direction due to the independence of perpendicular components of motion. For a float, the flow velocity $v(d_x, z)$ at its x-displacement d_x is time-stationary and dependent only on its depth z. The dynamics of a float is described in (Booth 1981):

$$m\ddot{d}_x = c \cdot \text{sign}(v(d_x, z) - \dot{d}_x) \cdot (v(d_x, z) - \dot{d}_x)^2, \tag{47}$$

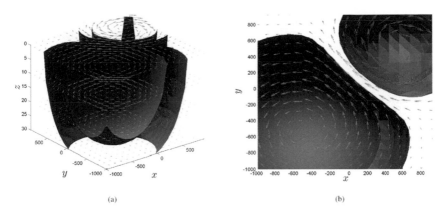

(a) (b)

Figure 2. (a) The three-dimensional flow field; (b) top view of the flow field.

where m is the constant rigid mass and c the drag parameter. The right hand side of the above equation represents the drag force that quantifies the resistance exercised on the profiling float in the flow field.

From (47), we define two state variables $x_1 := d_x$ and $x_2 := \dot{d}_x$. Further, $v(d_x, z)$ can be viewed as the unknown external input into the profiling float, implying the definition of $u := v(d_x, z)$. Then (47) can be rewritten as,

$$\dot{x}_1 = x_2,$$
$$\dot{x}_2 = \frac{c}{m} \cdot \text{sign}(u - x_2) \cdot (u - x_2)^2. \tag{48}$$

Its discrete-time representation, obtained by hypothetically holding the input constant over half open intervals $[kT, (k + 1)\,T)$, doing forward finite difference and measuring output measurements at kT, is given by

$$x_{1,k+1} = x_{1,k} + T \cdot x_{2,k},$$
$$x_{2,k+1} = x_{2,k} + T \cdot \frac{c}{m} \cdot \text{sign}(u_k - x_{2,k}) \cdot (u_k - x_{2,k})^2, \tag{49}$$

where $u_k := u(kT)$ and $x_{i,k} := x_i(kT)$ for $i = 1, 2$. The above equation can be expressed as

$$\mathbf{x}_{k+1} = \mathbf{f}(\mathbf{x}_k, u_k), \tag{50}$$

where $\mathbf{f}$ can be easily determined from the equations.

The motion of the float is characterized by an irregularly cycling submerging/surfacing pattern—it submerges and moves underwater for a certain duration, then resurfaces, and repeats the process over time. No matter whether it is underwater or on the surface, the depth $z_k := z(kT)$ and the acceleration $\ddot{d}_{x,k} := \ddot{d}_x(kT)$ are measurable; however, the position $d_{x,k} := d_x(kT)$ can only be measured when it is at surface. Thus, irregularly sampled measurements arise as a result, with the fast one $\tau_k := \ddot{d}_{x,k}$ and the slow one $\eta_k := d_{x,k}$ given by, respectively,

$$\tau_k = \frac{c}{m} \cdot \text{sign}(u_k - x_{2,k}) \cdot (u_k - x_{2,k})^2, \tag{51}$$

$$\eta_k = x_{1,k}.$$

For simplicity of notation, we rewrite (51) as

$$\tau_k = \varphi(u_k, x_k), \tag{52}$$

$$\eta_k = \phi(x_k).$$

Combining (49) and (51), we obtain the state space model to capture the dynamics of the float:

$$\Sigma : \begin{cases} \mathbf{x}_{k+1} = \mathbf{f}(\mathbf{x}_k, u_k) + \mathbf{w}_k, \\ \mathbf{y}_k = \mathbf{h}(\mathbf{x}_k, u_k) + \mathbf{v}_k, \end{cases} \tag{53}$$

Here, when the float is underwater, $\mathbf{y}_k = \tau_k$ and $\mathbf{h} = \varphi$; when at surface, $\mathbf{y}_k = \begin{bmatrix} \tau_k^T & \eta_k^T \end{bmatrix}^T$ and $\mathbf{h} = [\varphi^T \; \phi^T]^T$. In addition, $\mathbf{w}$ and $\mathbf{v}$ are added to account for noise in real world. They are assumed to be white Gaussian and independent of each other. The proposed SISF-wDF and SISS-wDF algorithms are applicable to the system Σ in (53) to acquire the information estimates of not only the velocities of the flow field (unknown input variables) but also the trajectory and velocity profile of the float (state variables).

5.2 Numerical Simulation

The flow field considered has dimensions of $(-1000, 1000)$ m $\times$ $(-1000, 1000)$ m $\times$ $(0, 30)$ m and the eddies are centered at $(500, 500)$ m and $(-500, -500)$ m, respectively, as shown in Figure 2. Compared to the typical size of these flows, its scale is intentionally narrowed to reduce computational burden (this does not restrict the applicability of the proposed algorithms to larger flow fields). Let 20 profiling floats be deployed evenly along the line segment from $(-800, -1000)$ m to $(1000, 800)$ m. The mass of a profiling float is 1.5 Kg, the drag coefficient c is 2 N $\cdot$ s^2/m^2, and the sampling period T is 0.05 s. The total traversing duration is 52 min.

The `SISF-wDF` and `SISS-wDF` algorithms are used together to build smoothed estimates of the inputs and states in the state-space model of a float. Let us examine the float released at (−400, −600) m and consider its motion in the x-direction. Figure 3 demonstrates the filtered and smoothed estimation of the flow velocity, y-displacement, and x-velocity respectively. We observe that the smoothed estimates are overall closer to the truth than the filtered ones. This is particularly evident for input estimation in Figure 3a—the filtered input estimates are quite noisy, but the smoothing reduces the errors significantly, thus improving the reconstruction accuracy. A quantitative comparison is further shown in Table 5. The metric is accumulative estimation error, defined by

$$\mathrm{Err}_s = \frac{1}{N}\sum_{k=1}^{N}(\hat{s}_k - s_k)^2.$$

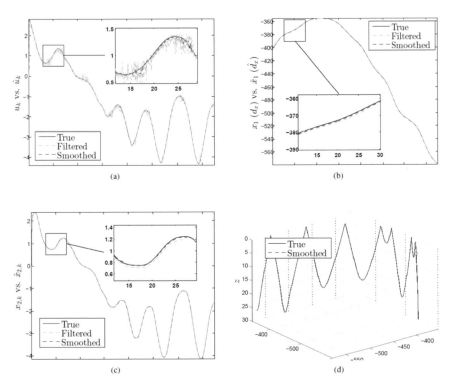

Figure 3. Estimation results for the profiling float released at (−400, −600) m: (a) u—flow velocity; (b) x_1—displacement along x-direction; (c) x_2—velocity along x-direction; (d) trajectory of the float (the circle denotes the location where the float is released).

Table 5. A quantified comparison between filtering and smoothing errors.

	Filtering	**Smoothing**
Err_u	340.02	90.61
Err_{x_1}	33931.97	19123.04
Err_{x_2}	83.08	26.89

where s_k for $k = 0,1,\ldots, N$ is a discrete-time signal and $\hat{s}_k$ is its estimate. Table 5 shows that the smoothing errors are considerably smaller than the filtering errors, illustrating the enhancement of accuracy achieved by smoothing. Figure 3d shows a good match between the smoothed trajectory and the true one.

Further, the estimated inputs of all profiling floats, which are the smoothed flow velocity data at different locations, are put together and used to reconstruct the flow field *via* the tessellation-based linear interpolation. The first and second rows in Figure 4 illustrate the true and reconstructed flow velocity fields along

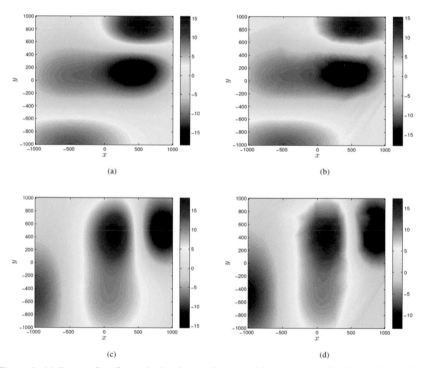

Figure 4. (a) True surface flow velocity along x-direction; (b) smoothed surface flow velocity along x-direction; (c) True surface flow velocity along y-direction; (b) smoothed surface flow velocity along y-direction.

x- and y-directions, respectively. One can see that the estimated velocities at the two eddies and the transition area are quit close to the actual case. Despite some minor differences observed, the overall reconstruction accuracy is quite satisfactory. This indicates that the proposed algorithms are able to provide reliable input and state estimates.

6. Conclusion

The world has increasingly realized the scientific and commercial importance of oceans for our society. Propelled by this trend, research of ocean monitoring based on advanced mechatronic systems has seen remarkable progress. This chapter considers the critical problem of ocean flow field estimation based on inexpensive profiling floats. The success of such an ocean observing system largely depends on the availability and effectiveness of algorithms capable of extracting necessary information from the motion data of the floats. Our analysis shows that the fundamental problem is to achieve joint input and state estimation of a dynamic system, which motivates our focus on the SISE problem of the dynamic systems from the perspective of Bayesian estimation, considering both the scenarios of filtering (SISF) and smoothing (SISS). We propose Bayesian paradigms, revealing how to update the probability density functions of the unknown input and state variables conditioned on the measurement data. Based on the paradigms and leveraging some Gaussian distribution assumptions, we formulate MAP estimation problems and exploit the Gauss-Newton method to build numerical solutions, leading to a catalog of SISF and SISS algorithms. We validate the effectiveness of our approach by applying two of the algorithms to simulations for flow field estimation using a swarm of profiling floats.

References

Anderson, B.D.O. and Moore, J.B. (1979). *Optimal Filtering*. Prentice-Hall, Englewood Cliffs.

Bell, B. and Cathey, F. (1993). The iterated Kalman filter update as a Gauss-Newton method. IEEE Transactions on Automatic Control, 38(2): 294–297.

Björck, Å. (1996). Numerical Methods for Least Squares Problems. SIAM, Philadelphia.

Booth, D.A. (1981). On the use of drogues for measuring subsurface ocean currents. Ocean Dynamics, 34: 284–294.

Candy, J.V. (2009). Bayesian Signal Processing: Classical, Modern and Particle Filtering Methods. Wiley-Interscience, New York, NY, USA.

Cheng, Y., Ye, H., Wang, Y. and Zhou, D. (2009). Unbiased minimum-variance state estimation for linear systems with unknown input. Automatica, 45(2): 485–491.

Colgan, C. (2006). Underwater laser shows. Explorations, Scripps Institution of Oceanography, 12(4): 20–27.

Corless, M. and Tu, J. (1998). State and input estimation for a class of uncertain systems. Automatica, 34(6): 757–764.

Costanza, R. (1999). The ecological, economic, and social importance of the oceans. Ecological Economics, 31: 199–213.

Darouach, M., Zasadzinski, M. and Boutayeb, M. (2003). Extension of minimum variance estimation for systems with unknown inputs. Automatica, 39(5): 867–876.

Darouach, M. and Zasadzinski, M. (1997). Unbiased minimum variance estimation for systems with unknown exogenous inputs. Automatica, 33(4): 717–719.

Fang, H., de Callafon, R.A. and Cortés, J. (2013). Simultaneous input and state estimation for nonlinear systems with applications to flow field estimation. Automatica, 49(9): 2805–2812.

Fang, H., de Callafon, R.A. and Franks, P.J.S. (2015). Smoothed estimation of unknown inputs and states in dynamic systems with application to oceanic flow field reconstruction. International Journal of Adaptive Control and Signal Processing, pp. 1224–1242.

Fang, H. and de Callafon, R.A. (2012). On the asymptotic stability of minimum-variance unbiased input and state estimation. Automatica, 48(12): 3183–3186.

Fang, H. and de Callafon, R.A. (2013). Simultaneous input and state smoothing and its application to oceanographic flow field reconstruction. In Proceedings of American Control Conference, pp. 4705–4710.

Fang, H. and de Callafon, R. (2011). Nonlinear simultaneous input and state estimation with application to flow field estimation. In Proc. of IEEE Conference on Decision and Control and European Control Conference (CDC-ECC), pp. 6013–6018.

Fang, H., Shi, Y. and Yi, J. (2008). A new algorithm for simultaneous input and state estimation. In Proceedings of American Control Conference, pp. 2421–2426.

Fang, H., Shi, Y. and Yi, J. (2011). On stable simultaneous input and state estimation for discrete-time linear systems. International Journal of Adaptive Control and Signal Processing, 25(8): 671–686.

Fang, H., Srivas, T., de Callafon, R.A. and Haile, M.A. (2017). Ensemble-based simultaneous input and state estimation for nonlinear dynamic systems with application to wildfire data assimilation. Control Engineering Practice, 63: 104–115.

Floquet, T., Edwards, C. and Spurgeon, S.K. (2007). On sliding mode observers for systems with unknown inputs. International Journal of Adaptive Control & Signal Processing, 21(8-9): 638–656.

Friedland, B. (1969). Treatment of bias in recursive filtering. IEEE Transactions on Automatic Control, 14(4): 359–367.

Gillijns, S. and De Moor, B. (2007a). Unbiased minimum-variance input and state estimation for linear discrete-time systems. Automatica, 43(1): 111–116.

Gillijns, S. and De Moor, B. (2007b). Unbiased minimum-variance input and state estimation for linear discrete-time systems with direct feedthrough. Automatica, 43(5): 934–937.

Gut, A. (2005). Probability: A Graduate Course. Springer, New York.

Han, Y., De Callafon, R.A., Cortés, J. and Jaffe, J. (2010). Dynamic modeling and pneumatic switching control of a submersible drogue. In International Conference on Informatics in Control, Automation and Robotics, 2: 89–97, Funchal, Madeira, Portugal.

Ha, Q.P. and Trinh, H. (2004). State and input simultaneous estimation for a class of nonlinear systems. Automatica, 40(10): 1779–1785.

Honerkamp, J. (1993). Stochastic Dynamical Systems: Concepts, Numerical Methods, Data Analysis. Wiley, New York, NY, USA.

Hsieh, C.-S. (2000). Robust two-stage Kalman filters for systems with unknown inputs. IEEE Transactions on Automatic Control, 45(12): 2374–2378.

Hsieh, C.-S. (2010). On the optimality of two-stage Kalman filtering for systems with unknown inputs. Asian Journal of Control, 12(4): 510–523.

Hsieh, C.-S. (2011). Optimal filtering for systems with unknown inputs via the descriptor Kalman filtering method. Automatica, 47(10): 2313–2318.

Hsieh, C.S. (2013). A unified framework for state estimation of nonlinear stochastic systems with unknown inputs. In Proceedings of the 9th Asian Control Conference, pp. 1–6.

Imine, H., Delanne, Y. and M'Sirdi, N.K. (2006). Road profile input estimation in vehicle dynamics simulation. Vehicle System Dynamics, 44(4): 285–303.

Kitanidist, P.K. (1987). Unbiased minimum-variance linear state estimation. Automatica, 23(6): 775–778.

Kitanidis, P.K. (1987). Unbiased minimum-variance linear state estimation. Automatica, 23(6): 775–778.

Mendel, J. (1977). White-noise estimators for seismic data processing in oil exploration. IEEE Transactions on Automatic Control, 22(5): 694–706.

Mutambara, A.G.O. (1998). Decentralized estimation and control for multisensor systems. CRC Press, Inc., Boca Raton, FL, USA.

Ouimet, M. and Cortes, J. (2014). Robust, distributed estimation of internal wave parameters via inter-drogue measurements. IEEE Transactions on Control Systems Technology, 22(3): 980–994.

Pina, L. and Botto, M.A. (2006). Simultaneous state and input estimation of hybrid systems with unknown inputs. Automatica, 42(5): 755–762.

Riser, S.C., Freeland, H.J., Roemmich, D., Wijffels, S., Troisi, A., Belbeoch, M., Gilbert, D., Xu, J., Pouliquen, S., Thresher, A., Le Traon, P.-Y., Maze, G., Klein, B., Ravichandran, M., Grant, F., Poulain, P.-M., Suga, T., Lim, B., Sterl, A., Sutton, P., Mork, K.-A., Vlez-Belch, P. J., Ansorge, I., King, B., Turton, J., Baringer, M. and Jayne, S.R. (2016). Fifteen years of ocean observations with the global Argo array. Nature Climate Change, 6(2): 145–153.

Robinson, E. (1957). Predictive decomposition of seismic traces. Geophysics, 22(4): 767–778.

Schubert, U., Kruger, U., Wozny, G. and Arellano-Garcia, H. (2012). Input reconstruction for statistical-based fault detection and isolation. AIChE Journal, 58(5): 1513–1523.

Shi, D., Chen, T. and Darouach, M. (2016). Event-based state estimation of linear dynamic systems with unknown exogenous inputs. Automatica, 69: 275–288.

Spinello, D. and Stilwell, D. (2010). Nonlinear estimation with state-dependent Gaussian observation noise. IEEE Transactions on Automatic Control, 55(6): 1358–1366.

Yong, S.Z., Zhu, M. and Frazzoli, E. (2014). Simultaneous input and state smoothing for linear discrete-time stochastic systems with unknown inputs. In Proceedings of IEEE Conference on Decision and Control, pp. 4204–4209.

Yong, S.Z., Zhu, M. and Frazzoli, E. (2016). A unified filter for simultaneous input and state estimation of linear discrete-time stochastic systems. Automatica, 63: 321–329.

You, F.-Q., Wang, F.-L. and Guan, S.-P. (2008). Hybrid estimation of state and input for linear discrete time-varying systems: A game theory approach. Acta Automatica Sinica, 34(6): 665–669.

4

Modern and Traditional Applications of Rheological Fluids in Mechatronic and Robotic Systems

Sylvester Sedem Djokoto,* Dragašius Egidijus, Vytautas Jurenas, Ramutis Bansevicius and *Shanker Ganesh Krishnamoorthy*

1. Introduction

Recently, Rheological Fluids (RFs) have been used in several applications related to mechanical and mechatronic fields due to their unique properties such as RFs are smart materials with controllable rheological properties. Also, it is important to understand the working principles and what mechatronics systems entails, which includes among others, sensors, actuators, signals, simulations and modeling. All these topics are explained in detail in this chapter according to the concepts and results reported in the literature. The chapter was divided into four parts, first part contains a brief description of mechatronics engineering, in particular definition, history, and different types. Whereas second part was focused on the rheological fluid and gives an overview of RFs (electrorheological fluid (ERF) and

Kaunas University of Technology, Studentu 56, LT-51424, Lithuania.
*Corresponding author: sedem.djok@gmail.com

magnetorheological fluid (MRF)). While the third part was focused on some modern application in mechatronics and robotic systems including; telerobotic surgery device that uses MRF, drive by wire cars, smart dampers in vehicles and other systems, smart clutches and the trunk robotic arm manipulators using rheological fluids in their joints for precision output. Finally, in the last part the authors suggested some future applications for ERF and MRF in cutting-edge technology

2. Mechatronic Engineering

2.1 Definition of Mechatronic

The term mechatronics was originally defined by the Yasakawa Electric Company. In trademark application documents, Yasakawa defined mechatronics in this way (Kyura and Oho 1996, Mori 1969). The word, mechatronics, is composed of "mecha" from the mechanism and the "tronics" from electronics. In other words, technologies and developed products will be incorporating electronics more and more into mechanisms, intimately and organically, and making it impossible to tell where one ends and the other begins.

The definition of mechatronics continued to evolve after Yasakawa suggested the original definition. One oft-quoted definition of mechatronics was presented by Harshama, Tomizuka, and Fukada in 1996 (Harshama et al. 1996). Further, mechatronics is the synergistic integration of mechanical engineering with electronics and intelligent computer control in the design and manufacturing of industrial products and processes.

Auslander and Kempf (1996) suggested another definition in that same year— mechatronics is the application of complex decision making to the operation of physical systems. Yet another definition by Shetty and Kolk (1997) appeared. They suggested that mechatronics is a methodology used for the optimal design of electromechanical products. More recently, Bolton (1999) found out that a mechatronic system is not just a marriage of electrical and mechanical systems and is more than just a control system; it is a complete integration of all of them.

All of these definitions and statements about mechatronics are accurate and informative, yet each one in and of itself fails to capture the totality of mechatronics. Despite continuing efforts to define mechatronics, to classify mechatronic products, and to develop a standard mechatronics curriculum, a consensus opinion on an all-encompassing description of "what is mechatronics" eludes us. This lack of consensus is a healthy sign. It says that the field is alive, that it is a youthful subject. Even without an unarguably definitive description of mechatronics, engineers understand from the definitions given above and from their own personal experiences the essence of the philosophy of mechatronics.

2.2 History of Mechatronic

For many practicing engineers on the front line of engineering design, mechatronics is not new. Many engineering products of the last 25 years integrated mechanical, electrical, and computer systems yet were designed by engineers that were never formally trained in mechatronics. Modern concurrent engineering design practices, now formally viewed as part of the mechatronics specialty, are natural design processes. What is evident is that the study of mechatronics provides a mechanism for scholars interested in understanding and explaining the engineering design process to define, classify, organize, and integrate many aspects of product design into a coherent package. As the historical divisions between mechanical, electrical, aerospace, chemical, civil, and computer engineering become less clearly defined, we should take comfort in the existence of mechatronics as a field of study in academia. The mechatronics specialty provides an educational path, that is, a roadmap, for engineering students studying within the traditional structure of most engineering colleges.

Mechatronics is generally recognized worldwide as a vibrant area of study. Undergraduate and graduate programs in mechatronic engineering are now offered in many universities. Refereed journals are being published and dedicated conferences are being organized and are, generally, highly attended. It should be understood that mechatronics is not just a convenient structure for investigative studies by academicians; it is a way of life in modern engineering practice. The introduction of the microprocessor in the early 1980s and the ever-increasing desired performance to cost ratio revolutionized the paradigm of engineering design. The number of new products being developed at the intersection of traditional disciplines of engineering, computer science, and the natural sciences is ever increasing. New developments in these traditional disciplines are being absorbed into mechatronics design at an ever-increasing pace. The ongoing information technology revolution, advances in wireless communication, smart sensors design (enabled by MEMS technology), and embedded systems engineering ensures that the engineering design paradigm will continue to evolve in the early twenty-first century.

2.3 Types of Mechatronic

The previous studies classified the mechatronic systems according to its application. The key elements of mechatronics are illustrated in Figure 1. As the field of mechatronics continues to mature, the list of relevant topics associated with the area will most certainly expand and evolve (Bishop 2002).

In recent years, RF has been a functional material used in mechatronic systems such as small and microscale transducers, sensors and/or actuators, and precision mechatronic control systems. It was not until the mid-1980s that scientists and

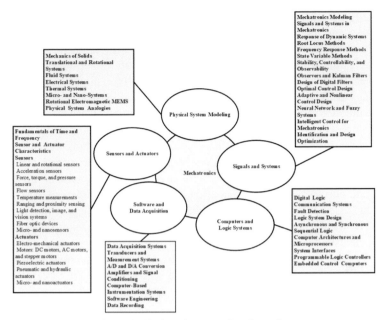

Figure 1. The key elements of mechatronics.

engineers started integrating RF materials with large-scale structures as *in situ* sensors and/or actuators, thus introducing the concept of smart materials such as RF, smart structures, and structronic systems. The next section gives an overview of RF and their applications.

3. Rheological Fluid (RF) and their Applications

3.1 History of RF

During the last few decades, rheological or smart fluids have attracted a significant amount of attention to their enormous potential in engineering applications such as mechatronics system. Smart fluids have the capability of both actuating and sensing when external input fields such as electric intensity or magnetic intensity are applied. Many studies have been conducted on smart fluids in several academy societies including chemistry, polymer, physics, and engineering. Recently, four different smart fluids, magnetorheological fluid (MRF) (De Vicente et al. 2011), electrorheological fluid (ERF) (Lee et al. 2001), magnetorheological elastomer (MRE) (Ju et al. 2012), and electroconjugate liquid (Seo et al. 2007) are actively being researched in various application fields. Their material characterizations are steadily investigated in terms of controllability of rheological properties such as viscosity and the field-dependent yield stress. In the last two decades, various research using smart fluids have been improved and the application technology using these smart fluids have been widely and rapidly undertaken by numerous researchers (Choi and Han 2012, Werely et al. 2008, Li et al. 2003, Choi and Han 2007, Sung et al. 2013, Carlson and Jolly 2000).

The inherent controllability of smart fluids has catalyzed broad-based research and development of many different systems including vehicle dampers, vibration control mounts, intelligent hydraulic systems, and smart robots. For the specific application, semi-active shock absorbers, utilizing MRF as the working fluid, have been successfully implemented on several passenger vehicles including Cadillac CTS-V and Ferrari FF. Researchers at the Intelligent Structures and Systems Lab (ISSL) and associates at the Ohio State University (OSU) medical school have been working on the development of novel haptic and force feedback devices aimed at improvements in telerobotic surgical systems using magnetorheological fluid. For minimally invasive cardio-thoracoscopic (MICT) surgery, the present state of the art is a telerobotic surgical system where the surgeon sits at the workstation and controls a robot, which conducts the surgery as are explained in the applications in Section 4.1.

3.2 Properties of RF

A rheological fluid is a fluid whose properties (for example viscosity, yield stress, surface tension, etc.) can be changed by applying a magnetic field or an electric field. Popular examples of smart fluids are electrorheological fluids (ERF) and magnetorheological fluids (MRF). The flow behavior of rheological fluid is like a Newtonian fluid, where the relationship between the shear stress and shear strain rate of the fluid is linear. By applying an electric/magnetic field to the smart fluid, the behavior of the fluid will be more similar to Bingham plastic fluid behavior as shown in Figure 2 (Spencer and Nagarajaiah 2003). To interpret this modification microscopically, the particles are spread within the rheological fluid without an electric/magnetic field; by applying an appropriate field, the particles are aligned into chains. Once aligned in this manner, the state of the fluid changes from a free-flowing liquid state to a solid-like state and the behavior of the fluid becomes more like Bingham plastic fluid as shown in Figure 3 (Elderaat 2013). As a result of applying an electric or magnetic field, a yield stress develops in the fluid. Rheological fluid is capable of developing yield stress in only a few milliseconds. This yield is a function of an applied field, and increasing an applied field will lead to increasing the yield stress. Figure 4 shows the change of yield stress and apparent viscosity by changing the applied field. The shear stress (τ) of the rheological fluid can be calculated by Equation 1 (Sims et al. 1999).

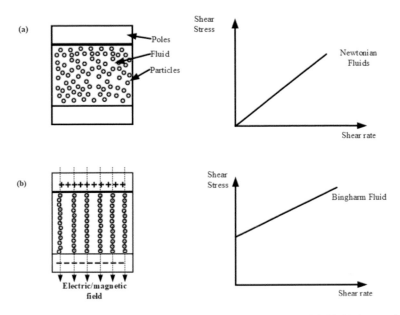

Figure 2. (a) Characterizing of rheological fluid without applying an external field, (b) characterizing of rheological fluid by applying an external field.

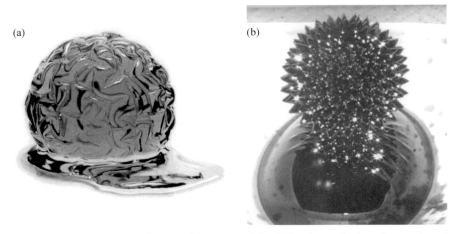

Figure 3. Optical microscope images of (a) MRF and (b) ERF under application of magnetic and electric fields respectively.

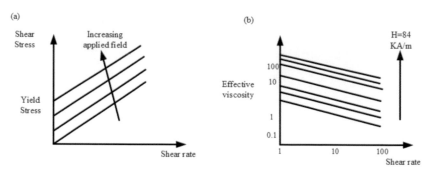

Figure 4. Variation of the shear stress and apparent viscosity with shear strain for an MR fluid under different magnetic field strengths: (a) Shear stress, and (b) the apparent viscosity (Wang and Liao 2011).

$$\tau = \tau_y(H)sgn(\dot{\gamma}) + \mu\dot{\gamma} \tag{1}$$

where τ_y, H, μ, and γ are defined as a yielding shear stress, applied field, viscosity of fluid, and shear strain rate.

The discovery and continuous research into smart fluids have in recent years offered solutions to many engineering challenges. There have been successes made by the use of smart fluids in many disciplines, ranging from the automotive and civil engineering communities to the biomedical engineering community. The research into this field of study has been impressive and Kaunas University of Technology (KTU) is part of the research institutions that have continual research studies into the identification and benefits of using rheological fluid devices in other areas apart from the ones mentioned above. The continuous research and documentation of these smart fluids continue to motivate current and future applications of these fluids and materials. Much of the success of rheological fluid devices is largely due to advancements in fluid technology.

3.3 Electrorheological Fluids (ERF)

Electrorheological fluids (ERF) are a suspension of extremely fine non-conducting polarizable particles (0.1–50 μm in diameter) in an electrically insulating fluid. An ER fluid undergoes an increase in viscosity upon application of an electric field. This was first reported by Winslow (1949) and is termed as "Winslow effect". This discovery leads to the development of ER-based various engineering applications such as breaks, clutches, damper, hydraulic valves, other applications such as bulletproof vests have been proposed for these fluids (Zhang et al. 2008). However, ER fluids have not found widespread commercial applications. This is mainly due to

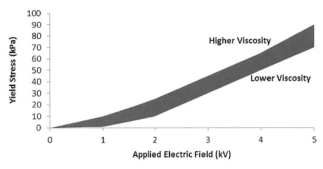

Figure 5. Yield Stress Performance of Gaint ERF (Smart Materials Laboratory Limited).

the fact that the maximum yield strength achieved by most of the conventional fluids is less than 15 kPa, much below 30 kPa required by many engineering devices and long-term use is not possible because of sedimentation problem (Huang et al. 2006).

Wen et al. (2003) discovered a new generation of ERF called giant ER (GER). The giant electrorheological (GER) fluid consists of the nanoparticles of oxalate core with urea coating and silicone oil. This fluid has high yield stresses (100–200 kPa) as shown in Figure 5. There are other merits of such GER fluid including fast response time, much slow sedimentation rate, and high breakdown voltage as reported in the application of the piezoelectric rotary motor (Qiu et al. 2014).

Shen et al. (2005) demonstrated that by adding a small amount of oleic acid to nanoparticles of barium titanyl oxalate coated with urea suspended in hydrocarbon oil produced a high yield stress. However, this dramatic increase of the dynamic yield stress also coincided with a sharp increased current density. In another research, Li et al. (2010) added oxide-carbon nanotube composites to the synthesis of urea-coated particles. The particles, when dispersed in different types of silicone oil, were shown to have enhanced the anti-sedimentation property. The yield stress, however, showed a 10% reduction (Shen et al. 2005).

3.4 Magnetorheological Fluids (MRF)

Magnetorheological fluids (MRFs) are classified as smart materials with controllable rheological properties. MR fluids are capable of achieving yield stresses in excess of 80 kPa which indicates an impressive range of fluid controllability and dynamic range. The fluids have also exhibited improved stability behavior. Moreover, the durability and life of the fluid have developed such that the fluid can be considered for commercial implementation. The performance of today's MR fluids is the result of a great number of studies which identify the properties and behavior of MR fluids. The literature is well populated with works related to the behavior of MR fluid or the performance of specific MR fluid devices (Carlson and Weiss 2004).

In each of these applications, the operating conditions of the fluid vary considerably. Perhaps a more extreme example would be the use of MR automotive dampers (intended for vehicle primary suspensions) in impact or shock loading applications. It is also used in civil engineering applications, for example, in bridges and building in earthquake zones. In medicine, it is used in prosthetic limbs and hands.

4. Applications of RF

4.1 Telerobotic Surgery

The researchers at the Intelligent Structures and Systems Lab (ISSL) and associates at the Ohio State University (OSU) medical school have been working on the development of novel haptic and force feedback devices aimed at improvements in telerobotic surgical systems. For minimally invasive cardio-thoracoscopic (MICT) surgery, the present state of the art is a telerobotic surgical system where the surgeon sits at workstation and controls a robot, which conducts the surgery. Some of the benefits of using telerobotic MICT surgery are smaller incisions, less pain, lower risk of infection, and less scarring. In addition, the recovery time for a person is reduced by at least a factor of 4 when minimally invasive surgery is used. In MICT surgery only three small incisions are needed and stopping the heart is optional see Figure 6.

In the same novelty of the minimally invasive telerobotic surgery system, other researchers have studied the use of smart materials in haptic systems. Recently, researchers have explored providing force feedback using ERF devices. Pfeiffer et al. (1999) have developed a 'smart glove' system named MEMICA (Mechanical Mirroring using Controlled stiffness and Actuators) using ER devices (Scilingo et al. 2000). The primary application area involves providing users with tactile

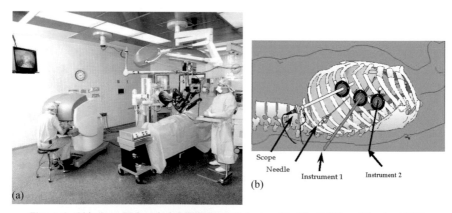

Figure 6. Ohio State University's MICT Surgical System (Da Vinci) (Ahmadkhanlou 2008).

feeling in a virtual reality environment. Sakaguchi et al. (2001) developed simple two-dimensional force display systems that demonstrate the use of ER devices in gaming applications. Two outcomes of the Sakaguchi work relate directly to this research. One is the fact that they had difficulty replicating anything but hard obstructions. The second lies in the discovery of the "sticky wall" phenomenon. This phenomenon occurs when one encounters a rigid obstruction and then wants to move away from it. Since the damper is in the "on" position, it doesn't move and the user feels the system being attracted or 'stuck' to the obstruction.

MR fluids, however, have certain advantages over their ER counterparts including an order of higher yield stress magnitude, broader operating temperature range, and lower sensitivity to impurities. These significant advantages make MR fluids more attractive in force feedback applications. Another major factor in favor of MR fluids is that ER fluids require very high voltages (on the order of a few kilovolts). This makes it hazardous to use ER fluids in force feedback systems where the user is in close contact with the device in operation. Scilingo et al. (2000) developed simple haptic devices using MR fluids to mimic the compressional compliance of different biological tissues and Jolly et al. (1998) developed MR based haptic devices. However, neither addressed the issue of transparency (Pfeiffer et al. 1999).

4.2 Drive-By-Wire

Another novel application of rheological fluids is in the area of automobile design. Recently, the idea of replacing the hydraulic and mechanical systems of automobiles by proven aerospace technology has been taken under consideration. Systems that operate automobiles by means of computer-controlled electronic signals instead of direct action on control devices are called drive-by-wire systems. The term sounds like fly-by-wire, which is a method of controlling commercial aircraft that has been in use for more than a decade. For more than a century, drivers controlled vehicles under direct actuation by mechanical, hydraulic, or pneumatic systems. If such a system fails, the consequences could affect the safety of the vehicle or passengers. There is considerable interest in increasing functionality and safety by developing drive-by-wire systems where electronic controls are used to supplement the driver controls. In a drive-by-wire system, the driver controls are simply inputs to a computerized system rather than directly commanding the vehicle functions. The idea is to remove the mechanical linkages between the controls of a car and replace it with electronic devices. Instead of operating the steering, accelerator, and brake directly, the electronic systems will send commands to a central computer, which will control the vehicle (Figure 7).

As operational information is conveyed by means of electronic signals, there is no mechanism for information feedback to the driver on actual road-surface

Figure 7. The Hy-wire's X-drive, driver's control unit, Photo courtesy General Motors (Pfeiffer et al. 1999).

conditions, vehicle status, and cautionary information. This is the most pertinent problem with computer control of steering systems. Such deficiency could be eliminated by capabilities of haptic technology. By employing force feedback technology, the drive-by-wire system conveys required information to the driver in tactile form. Feedback would certainly be required in order to give the driver meaningful information about what is happening at the road wheels. For instance, feedback on irregularities in the surface of an unpaved road is transmitted via movement of the steering wheel. Furthermore, the force feedback technology allows the communication of cautionary information to the driver through vibration or cessation of operation such as a warning when the driver is deviating from the lane, is following too closely, or has become drowsy. Sensors are required for steering position and velocity both at the steering wheel and the road wheels. Torque measurements are required for the road wheels as well as the steering wheel when force feedback is applied as shown in Figure 8a.

For this application of MR dampers, we have the master (steering wheel) in the real world, while the slave (vehicle) is in a virtual reality environment. The steering wheel has an encoder measuring the steering wheel angle. For the sake of redundancy, another revolving digital sensor could be used. The MR damper is connected to the steering wheel shaft to give some feedback to the driver. By applying current to the MR damper (Figure 8b), it is possible to generate a resistance torque when the driver rotates the steering wheel. The torque is dependent on the steering wheel angle or the difference between the steering wheel angle and the angle of vehicle wheels. There should also be a speed dependent behavior, restricting the turning angle at high speed. There is an encoder on steering wheel shaft to measure the angle of rotation.

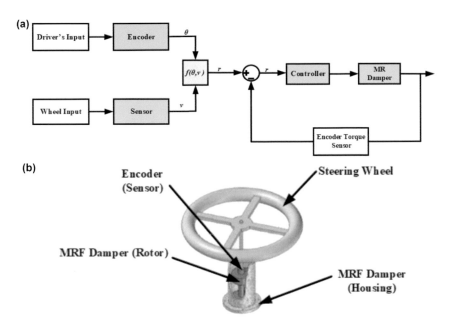

Figure 8. (a) Steer-by-wire control system and (b) Steering wheel components (Ahmadkhanlou 2008, Pfeiffer 1999).

4.3 Electrorheological Fluid Valve for Control of ERF Actuators

In recent studies, Sheng and Wen (2012) used ERFs as valves and micro-valves. These studies confirmed the basic functionality of using an ERF as a valve, though there is still a lack of empirical study in the literature that utilizes the ERF valve as part of an overall system in which nonlinear effects, such as particle distribution temperature and ERF particle purity, can play a major role in the performance of an ERF valve and are only observable through experimentation.

Figure 9 shows the assembled ERF valve connected to the ERF test-bed. The experiment measures the ERF's pressure signals, temperature, as well as the leak current that the ERF bonds conduct within the electrodes when there is an E-field. Two sensors capture pressure signals across the ERF valve. A Ponreed PTD504 absolute pressure sensor is placed in front of the valve and measures the valve input pressure, and a Ponreed PTD508B differential pressure sensor measures the pressure between the input and output ports of ERF valve. Motor speed control and data acquisition was done via a computer assembly model. The control computer

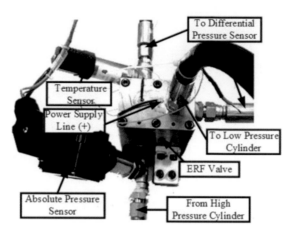

Figure 9. A close-up view of the assembled ERF valve in the test-bed (Nguyen et al. 2015).

uses a WEINVIEW TK6070IH as the human-machine interface of the system to show data and input motor control parameters; a KEYENCE KV-N60AT as the programmable logic controller to control the motor driver and a KV-NC4AD as the analog-digital converter for sending commands and retrieving sensor data. The power supply line showed in Figure 9 and the adjustable power supply provides the voltage across the ERF valve to generate the desired E-field and measures the ERF valve's leak current. The leak current helps monitor the arcing possibility in the ERF system by signaling the power supply to deactivate the E-field when the leak current is over 7.5 mA, thus protecting the ERF.

4.4 MRF-Safety-Clutch

There is a huge potential for technical application of magnetorheological fluids (MRF), beside adjustable dampers, switchable engines and so called smart clutches. There are two common designs of these clutches, the disc- and the bell-design. Both consist of two rotatable parts, the input or drive side and the output or power take-off side, with a small gap in between. This gap is filled with MRF that transmits the torsion moment from the input to the output side depending on its viscosity and its interaction with the surfaces. These concepts allow influencing the power transmission and realize an adjustable slip torque. Figure 10 shows a CAD model of the clutch.

4.5 MRF and ERF Dampers

Mechatronic suspension systems can ease the conflict of the objectives ride comfort, ride safety, and limited suspension deflection. In the last years, fully active

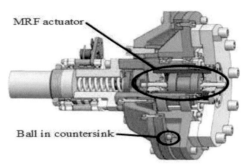

Figure 10. CAD Model of the MRF safety clutch (Jackel et al. 2013).

suspension system has been intensively studied. However, their high costs and high energy demand have limited their use. In production vehicles, semi-active dampers are primarily integrated which offer performance advantages over passive devices. They are mainly controlled by skyhook based comfort or road-holding oriented control laws using, in general, static force-velocity characteristics to determine the damper control inputs to track reference forces (Karnopp 1983). Due to the complex hardware structure, that involves not only the internal valve assemblies but also the switching external valves, the force/velocity response is highly nonlinear. It is challenging to develop an effective control strategy for these road conditions. In order to fully exploit the potential of fast modern semi-active dampers, it is desired to employ model-based control techniques to incorporate hysteresis effects, valve

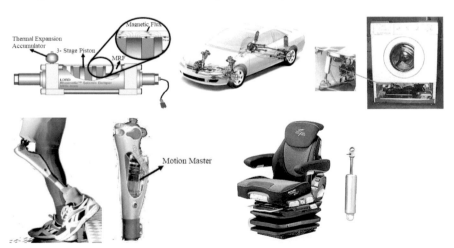

Figure 11. Examples of some ERF and MRF dampers in some mechatronics applications (Wang and Liao 2011).

81

dynamics, and dynamics of the damper force generation in the controller design. By the control approach, the nonlinear damper effects and the dynamics of the generation of the damper force are taken into account. Since the damper force is, in general, not measured, a model based feedforward control strategy for the damper current is designed. A skyhook based controller and a linear quadratic regulator (LQR), the artificial neural network could be, for example, used for the control. The measurement data is analyzed to evaluate the performance of the proposed approach compared to the static damper characteristics.

Where; Ms, Mu, Ks, Kt, and Ct are Sprung mass, Unsprung mass, Spring stiffness, Tire stiffness, Damping coefficient.

4.6 MRF and ERF Elephant Trunk Robotic Manipulator

Robot manipulators were first introduced by General Motors in 1961 (Carneige Mellon University 2014) for mass manufacturing. The manufacturing, mining, and oil exploration industries have benefited immensely from the introduction of robotic manipulators. These benefits are: efficient production of goods; and handling and controlling objects where it is inaccessible for humans, e.g., in subsea oil exploration where ROVs are used to access and control subsea equipment. These robotic manipulators could easily perform a great variety of tasks with speed, precision, and repeatability. Despite these benefits and many years of research, robots have still failed to become ubiquitous within human homes. This failure can be attributed to a large number of causes such as high cost, safety, lack of software intelligence, and the motion limitations of traditional manipulators. Figures 13a and b, 14 shows the traditional manipulator of the ROV and Space Station Remote Manipulator System respectively.

The robotic arm is normally designed with a large number of rigid links connected by rotational or prismatic joints. These rigid links can place motion constraints on these manipulators, which can make it difficult for the arm to reach certain positions or closely follow complex trajectories. Even when the manipulator is capable of reaching every desired point within the defined work area, joint limits and singularities may prevent it from doing so along a continuous path, making strict trajectory-following difficult. Often positioning a workpiece in an environment such that a specific trajectory is valid requires a trial and error methodology where one tries moving the workpiece into several possible poses until one is found which allows the robot to generate a desirable path. Obviously, such a methodology is not feasible outside of a choreographed industrial environment, making it difficult for average people to use. Making matters worse, when working in complex, obstacle-ridden environments, objects can easily obstruct otherwise valid trajectories. Even when a valid trajectory can be found, it may require placing the robot in awkward positions where it is unable to take full advantage of the

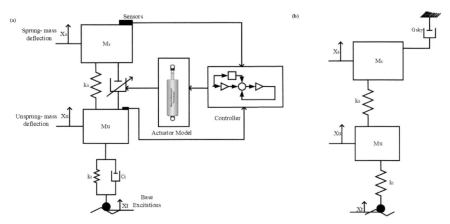

Figure 12. (a) Quarter-car semi-active model, (b) with Skyhook control strategy (Unsal 2006).

Figure 13. (a) Remotely Operated Vehicle and (b) manipulator for ROV (ROV Innovations).

available torque. Additionally, today's manipulators are often heavy, bulky, and dangerous, and are usually separated from humans by protective fences to limit human contact as much as possible for fear of accidents or fatalities.

To approach this problem, the Institute of Mechatronics at Kaunas University of Technology (KTU) has conducted research into the construction of a manipulator with significantly fewer motion constraints which could operate safely around humans and reach difficult places that the traditional robotic arm could not reach. Having fewer motion constraints means less intelligence will be needed to find valid and safe trajectories for the manipulator to move through, opening up opportunities for robotic manipulators to move into these environments.

The Institute of Mechatronics at Kaunas University of Technology under the SmartTrunk project (Grant for Research from Lithuanian Science Council

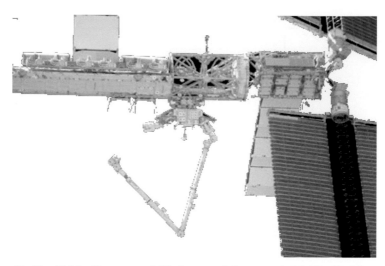

Figure 14. The Mobile Transporter (MT) is part of the outpost's Mobile Servicing System (MSS) which includes systems such as the Space Station Remote Manipulator System (SSRMS), called Canadarm 2, and the Special Purpose Dexterous Manipulator (SPDM), called Dextre (SPF SPACEFLIGHT INSIDER).

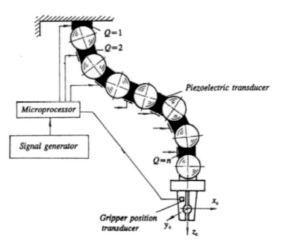

Figure 15. Trunk robotic arm with piezo actuator (Augustaitis and Kulvietis 2017).

2015) have been working on the development of using rheological fluids to solve problem associated with actuating the smart trunk like robot with different joint (Figure 15) (Grant for Research from Lithuanian Science Council 2015). Actuation is a technical term used in robotics and control theory to describe mechanical systems that cannot be commanded to follow arbitrary trajectories in configuration space (Surdhar and White 2003). This condition can occur due to a number of reasons, the simplest of which is when the system has a lower number of actuators than degrees of freedom. In this case, the system is said to be trivially underactuated. The under actuation manipulators are those which have more degrees of freedom than the number of actuators. They allow us to develop the simplest devices with reduced weight, cost, and energy consumption. The main problem is in organizing the control of the kinematic chain, consisting of both passive and active links (Bansevicius et al. 2007, Raphael and Oliver 2016, Luca De et al. 2001, Tso et al. 2003).

The number of DOF was increased, enabling the grip to reach hardly accessible zones. In this type of manipulator, there is a highly reduced number of actuators (e.g., electric motors) (Jain and Rodriguez 1993). It is a very simplified mechanical design and its low weight could be of great importance for satellite applications and soft robotic spines in subsea oil exploration which are highly flexible soft robotic arms and safe for human-machine interaction. The trajectories of grip are realized by applying special software, controlling in real-time the number of DOF of every kinematic pair in relation to the phase of the external force with alternating direction (as in the case when the centrifugal force is generated by a rotating unbalance).

The trunk robotic arm was modified from the previous experiment of using single 1D or 2D vibrator and piezoelectric control of DOF of kinematic pairs. Instead of using the centrifugal force of the rotating unbalanced rotor, a 1 DOF and 2 DOF multilayer bimorph exciters was applied as shown in Figure 15.

One modification to the trunk robotic arm is the application of Electrorheological Fluids (ERF) to the actuators of the trunk robotics arm shown in Figure 16. The first modification (Figure 16a) uses a rotating unbalance exciting force, where the

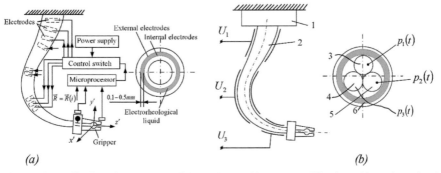

(a)   *(b)*

Figure 16. Application of ERF to control the trunk's position: two modifications: (a) – schematic of trunk robot with internal and external electrodes and the gap between them (0.1 ... 0.5 mm) filled with ERF; (b) – schematic of trunk robot excited by pressures with different phases: 1–3 phase pump; 2 – elastic element; 3, 4, 5 – elastic hoses; 6 – the layer of ERF (Bansevicius et al. 2007).

viscosity of specific parts of an elastic trunk is controlled by an electric field and is connected to electrodes with the help of a microprocessor and control switch which depend on the trajectory of the grip. The other modification is shown in Figure 16b uses a 3-phase pump (1) and to generate pressure in three elastic hoses (3, 4 and 5) that are placed in an elastic element (2). The pressure can be expressed as; $p_1(t) = p_0 \cos \Omega t$; $p_2(t) = p_0 \cos(\Omega t | 2\pi/3)$; $p_3(t) = p_0 \cos(\Omega t + 4\pi/3)$. The difference of phases in pressures p_i, where i = 1, 2, 3, causes the grip to oscillate in circles, except in the areas where the ERF, controlled by the electric fields Ui, where i = 1, 2, 3 (Figure 16b), has maximum viscosity. As in Figure 16a, the frequency Ω should be coordinated with the response time of the solidification of the ERF that is controlled by the electric fields Ui. Similar schematics can be shown using magnetorheological fluids (MRF). When the MRF is subjected to a magnetic field, the fluid increases its viscosity to the point of becoming a viscoelastic solid. Importantly, the yield stress of the fluid when in its active ("on") state can be controlled very accurately by varying the magnetic field intensity. It should be noted that it is much easier to generate an electric field than a magnetic field, but in some cases, where dimensions are not important, MRF is quite applicable.

5. Future Applications of RF

In the future, growth in use RF in mechatronics, robotics, and another mechanical system will be fueled by the perfection of the properties of these fluids. The shortfall in the traditional fluids such as hydraulic fluids gave rise to the use of RF in "enabling technologies". For example, the invention of the RF trunk robotic arm had a profound effect on the redesign of traditional robotic arm and design of new robotic systems. It is expected that this will give advancements in cost-effective actuator development enabled by advancements in applications of MEMS, adaptive, and control methodologies. The continued rapid development of Mechatronics and robotics will only accelerate the pace of smart product development. Developments in space vehicles and devices provide vivid examples of mechatronics development. There are also numerous examples of using underwater manipulators for subsea oil exploration as shown in Figure 17 (Kelasidi 2014). Other areas that will benefit from the use RF are Intelligent systems in all walks of life: including smart home appliances such as dishwashers, washing machines; agriculture machinery such as combine harvesters, etc.; automobile such as smart dampers, steering systems, etc.; entertainment; healthcare that includes hospitals, patient-care, surgery (robot-assisted surgery), research, etc.; laboratories of science, engineering, etc.; manufacturing; military; mining; excavation, exploration; transportation that includes air, ground, rail, space, etc.; microwaves; and wireless network-enabled devices. The future of RF in mechatronics and robotics is immense.

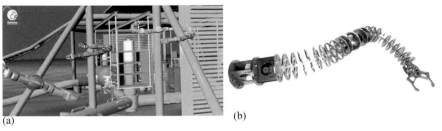

Figure 17. Swimming manipulators, an example of application of trunk robotic arm, this can perform their own locomotion as well as manipulation and intervention tasks (Kelasidi et al. 2014).

Conclusion and Outlook

In this chapter, the authors have presented the traditional and new applications of MRFs in mechatronics and robotic systems as was reported in the literature. In addition, the authors suggested that some future applications could be more promising in high-Tech applications such as aerospace. Based on the results mentioned in the literature, the MRFs and ERFs are widely used in many applications rated to mechatronics and robotic systems. In order to face the repaid development in these applications, the trend is to other additives such as (a polar molecule) to improve the behaviors of ERFs in terms of yield stresses under an electric field.

The typical knowledge base for the optimal design and operation of mechatronics and robotics using RF comprises of system modeling and analysis, decision and control theory, sensors and signal conditioning, and actuators and power electronics. The relevant technology applications of RF in mechatronics systems include medical, defense, manufacturing, robotics, and automotive.

References

Abhinandan Jain and Guillermo Rodriguez. (1993). An analysis of the kinematics and dynamics of underactuated manipulators. IEEE Trans Robot Autom, 9: 411–22.

Aistis Augustaitis and Genadijus Kulvietis. (2017). Kinematic of trunk-like robots using D-H homogenous transformation. Proceedings of 22nd International Conference MECHANIKA.

Auslander, D.M. and Kempf, C.J. (1996). Mechatronics: Mechanical System Interfacing, Prentice-Hall, Upper Saddle River, NJ.

Bansevicius, R., Parkin, R., Jebb, A. and Knight, J. (1996). Piezomechanics as a sub-system of mechatronics: present state of the art, problems, future developments. IEEETrans Ind. Electron, 43(1): 23–9.

Bansevicius, R., Sarkauskas, K.K., Tolockas. (2007). Rt. Underactuated manipulator with control based on variable dynamic properties of joints. Electron Electr. Eng., (7(79)) (Automation, Robotics, T125) ISSN 1392-1215.

Bishop, R.H. (2002). The Mechatronics Handbook, CRC Press, Florida USA, pp. 19.

Bolton, W. (1999). Mechatronics: Electrical Control Systems in Mechanical and Electrical Engineering, 2nd Ed. Addison-Wesley Longman, Harlow, England.

Carlson, J.D. and Weiss, K.D. (1994). A growing attraction to magnetic fluids. Machine Design, 66(15): 61–64.

Carlson, J.D. and Jolly, M.R. (2000). MR fluid, foam and elastomer devices. Mechatronics, 10(4): 555–569. Google Scholar CrossRef.

Carnegie Mellon University. Robot Hall Of Fame - Unimate, 2014. <http://www.robothalloffame.org/inductees/03inductees/unimate.html>.

Choi, S.-B. and Han, Y.-M. (2007). Vibration control of electrorheological seat suspension with human-body model using sliding mode control. Journal of Sound and Vibration, 303(1-2): 391–404. Google Scholar CrossRef.

Choi, S.B. and Han, Y.M. (2012). Magnetorheological Fluid Technology: Applications in Vehicle Systems, Taylor Francis, CRC Press, Boca Raton, Fla, USA. Google Scholar CrossRef.

De Luca, A., Iannitti, S., Mattone, R. and Oriolo, G. (2001). Advanced intelligent mechatronics. Proceedings. 2001 IEEE/ASME International Conference, 2: 855–61. ISBN: 0-7803-673607.

De Vicente, J., Klingenberg, D.J. and Hidalgo-Alvarez, R. (2011). Magnetorheological fluids: a review. Soft Matter, 7(8): 3701–3710, Google Scholar CrossRef.

Deimel Raphael, Brock Oliver. A novel type of compliant and underactuated robotic hand for dexterous grasping. Int. J. Robot Res. 2016 (January–March). Online ISSN: 1741–3176.

Eleni Kelasidi, Kristin Ytterstad Pettersen, Jan Tommy Gravdahl and Pål Liljebäck. (2014). Modeling of underwater snake robots. Robotics and Automation (ICRA), 2014 IEEE International Conference, 4540–4547.

Farzad Ahmadkhanlou. (2008). Design, Modeling, and Control of Magnetorheological Fluid-Based Force Feedback Dampers for Telerobotic Systems, Dissertation Presented in Partial Fulfillment of the Requirements for the Degree Doctor of Philosophy in the Graduate: School of The Ohio State University.

Grant for research from Lithuanian Science Council MIP 084-2015.

Haithem Ismail Elderrat. Research Towards The Design of A Novel Smart Fluid Damper Using A Mckibben Actuator, Thesis Submitted for the degree of Master of Philosophy: The University of Sheffield, June 2013, p. 3.

Harshama, F., Tomizuka, M. and Fukuda, T. (1996). Mechatronics—What is it, why, and how?—an editorial. IEEE/ASME Transactions on Mechatronics, 1(1): 1–4.

Jackel, M., Kloepfer, J., Matthias, M. and Seipel, B. (2013). The novel MRF-ball-clutch design – a MRF-safety-clutch for high torque applications, 13th Int. Conf. on Electrorheological Fluids and Magnetorheological Suspensions (ERMR2012) IOP Publishing. Journal of Physics: Conference Series, 412, 012051.

Jiaxing, Li, Xiuqing, Gong, Shuyu, Chen, Weijia, Wen and Ping, Sheng. (2010). Giant electrorheological fluid comprising nanoparticles: Carbon nanotube composite. Journal of Applied Physics 107: 093507.

Jolly, M.R., Carlson, J.D. and Bender, J.W. (1998). Properties and applications of commercial magnetorheological fluids. In SPIE 5th Annual Int. Symposium on Smart Structures and Materials, San Diego, CA.

Ju, B.X., Yu, M., Fu, J., Yang, Q., Liu, X.Q. and Zheng, X. (2012). A novel porous magnetorheological elastomer: preparation and evaluation. Smart Materials and Structures, 21(3), Article ID 035001. Google Scholar CrossRef.

Karnopp, D.C. (1983). Active damping in road vehicle suspension systems. Vehicle System Dynamics, 12: 291–316.

Kyura, N. and Oho, H. (1996). Mechatronics—an industrial perspective. IEEE/ASME Transactions on Mechatronics, 1(1): 10–15.

Lee, H.G., Choi, S.B., Han, S.S., Kim, J.H. and Suh, M.S. (2001). Bingham and response characteristics of ER fluids in shear and flow modes. International Journal of Modern Physics B, 15(6-7): 1017–1024. Google Scholar CrossRef.

Li, W.H., Du, H. and Guo, N.Q. (2003). Finite element analysis and simulation evaluation of a magnetorheological valve. International Journal of Advanced Manufacturing Technology, 21(6): 438–445. Google Scholar CrossRef.

Memet Unsal. (2006). Semi-Active Vibration Control of Parallel Platform Mechanism Using Magnetorheological Damping, Ph.D. dissertation, University of Florida. Mechanical Engineering Department, Gainesville, Florida.

Mori, T. (1969). Mechatronics, Yasakawa Internal Trademark Application Memo 21.131.01, July 12, 1969.

Pfeiffer, C., Mavroidis, C., Bar-Cohen, Y. and Dolgin, B. (1999). Electrorheological Fluid Based Force Feedback Device, in the 1999 SPIE Telemanipulator and Telepresence Technologies VI Conference, Boston, MA, pp. 88–99.

Qiu, W., Hong, Y., Mizuno, Y. et al. (2014). Non-contact piezoelectric rotary motor modulated by giant electrorheological fluid. Sensors and Actuators A: Physical, 217: 124–128.

Quang-Anh Nguyen, Steven Jens Jorgensen, Joseph Ho and Luis Sentis. (2015). Characterization and testing of an electrorheological fluid valve for control of ERF actuators. Journal Actuators, pp. 135–155.

ROV Innovations http://www.rovinnovations.com/manipulator-arms.html.

Sakaguchi, M., Furusho, J. and Takesue, N. (2001). Passive Force Display Using ER Brakes and its Control Experiments. In Virtual Reality Conference (VR'01), Yokohama, Japan.

Scilingo, E.P., Bicchi, A., Rossi, D.D. and Scotto, A. (2000). A magnetorheological fluid as a haptic display to replicate perceived compliance of biological tissues, in IEEE-EMBS Conference on Microtechnologies in Medicine & Biology, Lyon, France, pp. 229233.

Seo, W.-S., Yoshida, K., Yokota, S. and Edamura, K. (2007). A high-performance planar pump using electro-conjugate fluid with improved electrode patterns. Sensors and Actuators A: Physical, 134(2): 606–614. Google Scholar CrossRef.

Shen, R., Wang, X., Wen, W. and Lu, K. (2005). TiO based electrorheological fluid with high yield stress. International Journal of Modern Physics B, 19(7, 8 and 9): 1104–1109..

Sheng, P. and Wen, W. (2012). Electrorheological fluids: mechanisms, dynamics, and microfluidics applications. Annu. Rev. Fluid Mechan. 44: 143–174.

Shetty, D. and Kolk, R.A. (1997). Mechatronic System Design, PWS Publishing Company, Boston, MA.

Sims, N., Stanway, R. and Johnson, A. (1999). Vibration control using smart fluids: A state-of-the-art review. The Shock and Vibration Digest, 31(3): 195–203.

Smart Materials Laboratory Limited, 8/F., Houtex Industrial Building, 16 Hung To Road, Kwun Tong, Kowloon, Hong Kong, http://www.smartmaterials.hk/.

Spencer, B.F. and Nagarajaiah, S. (2003). State of the art of structural control. Journal of Structural Engineering, 129(7): 12.

Spf Spaceflight Insider http://www.spaceflightinsider.com/missions/iss/unscheduled-spacewalk-planned-space-station-crew-monday.

Sung, K.-G., Seong, M.-S. and Choi, S.-B. (2013). Performance evaluation of electronic control suspension featuring vehicle ER dampers. Meccanica, 48(1): 121–134. Google Scholar CrossRef.

Surdhar, J. and White, S. (2003). A parallel fuzzy-controlled flexible manipulator using optical tip feedback. Robot Comput-Integr Manuf, 19(3): 273–82.

Tso, S.K., Yang, T.W., Xu, W.L. and Sun, Z.Q. (2003). Vibration control for a flexible link robot arm with deflection feedback. Int. J. Nonlinear Mech., 38: 51–62.

Wang, D.H. and Liao, W.H. (2011). Magnetorheological fluid dampers: a review of parametric modeling. Smart Materials and Structures.

Wen, W.J., Huang, X.X., Yang, S.H., Lu, K.Q. and Sheng, P. (2003). Nature Materials 2: 727.

Wereley, N.M., Cho, J.U., Choi, Y.T. and Choi, S.B. (2008). Magnetorheological dampers in shear mode. Smart Materials and Structures, 17(1), Article ID 015022. Google Scholar CrossRef.

Winslow, W.M. (1949). Journal of Applied Physics, 11, 319.

Xianxiang Huang, Weijia Wen, Shihe Yang and Ping Sheng. (2006). Mechanism of the giant electrorheological effect. Solid State Communication, 139: 581–588.

Zhang, X.W., Zhang, C.B. and Yu, T.X. (2008). Characterization of electro-rheological fluids under high shear rate in parallel ducts. International Journal of Modern Physics B, 22(31 & 32): 6029–6036.

5

Parameter Identification and Damage Detection of Offshore Structural Systems

Jian Zhang,[1,]* *Xiaomei Wang*[2] and *Chan Ghee Koh*[3]

1. Introduction

System identification is an inverse analysis of the dynamic system to identify system parameters and detect parameter changes based on given input and output (I/O) information, in which three basic components are input excitation, dynamic system, and output response. In structural engineering, system identification is generally applied to structural parameter identification, damage detection, and health monitoring. Stiffness and damping of the dynamic system can be identified to update or calibrate the numerical model so as to better predict structural response and build cost-effective engineering structures. Furthermore, system identification methods can potentially be developed as a useful non-destructive evaluation method

[1] Faculty of Civil Engineering and Mechanics, Jiangsu University, Zhenjiang 212013, P.R. China.
[2] Department of Ocean Engineering, Tianjin University, Tianjin 300072, P.R. China.
[3] Department of Civil and Environmental Engineering, National University of Singapore, 1 Engineering Drive 2, Singapore 117576.
* Corresponding author: jianzhang@ujs.edu.cn

and can provide an in-service condition assessment or health monitoring of existing and retrofitted structures. In earlier days, only visual inspection by UAV or ROV and local non-destructive techniques such as ultrasound detection and acoustic emission method are available for structural health monitoring. However, visual inspection is often incomplete and local non-destructive techniques are limited to detection of individual structural components. In this regard, implementation of identification methods is able to globally and quantitatively identify the structural dynamic system as a real-time strategy. Numerous system identification methods have been developed including classical and non-classical methods and their brief introduction is given in this section.

System identification methods have been widely applied to onshore structures such as buildings and bridges. As demands for offshore exploration and production increase, it is necessary to extend the implementation of system identification to offshore engineering field to provide operators useful and timely information to detect adverse changes and present structural failure. Accidents and injuries are very common on offshore platforms. For example, more than 60 offshore workers died and more than 1,500 suffered injuries in offshore energy explorations and production in the Gulf of Mexico from 2001 through 2009, according to the data from the U.S. Minerals Management Service. Considering the relatively small cost compared to the total project cost and serious consequence of any undesired event, it is clear that system identification of offshore structures is highly beneficial and necessary to apply. However, compared to onshore structures, offshore structures are more complex dynamic systems in ocean environments and thus present more challenges for system identification. There are many uncertainties such as water-structure interaction and unknown initial conditions for the structural response, which make system identification more difficult due to the ill-conditioned nature of inverse analysis. Besides, environmental loads, especially random wave forces, are difficult to determine or measure in practice. It is, therefore, necessary to develop robust and effective strategies for parameter identification and damage detection of offshore structural systems.

Due to better cost-effectiveness and mobility, jack-up platforms have been installed and operated from initially shallow waters to deeper waters recently, where harsher environmental conditions are involved. In order to provide accurate safety assessment and early identification of potential damage, system identification of jack-up platform is highly recommended to develop and apply. Therefore, the jack-up platform is taken as an illustration example to study in this research.

1.1 System Identification Methods

A considerable number of reviews on system identification methods have been published such as Ghanem and Shinozuka 1995, Ewins 2000, Maia and Silva

2001, Chang et al. 2003, Carden and Fanning 2004, Hsieh et al. 2006, Humar et al. 2006, Friswell 2007, Fan and Qiao 2011, Dessi and Carmerlengo 2015. Based on different criteria for classification, system identification methods can be categorized into parametric and nonparametric models, deterministic and stochastic methods, frequency domain and time domain methods, and classical and non-classical methods. In this study, the last classification is used.

Classical Methods: Classical methods are usually derived from sound mathematical principles, e.g., filter methods, least squares methods, instrumental variable method, gradient search methods, maximum likelihood method, natural frequency-based method, and mode shape based methods. Although these methods have been successfully applied to many problems, the limitations are the requirement of a good initial guess, the sensitivity to noises and high probability of convergence to local optima in structural identification which normally involves many unknown parameters. Moreover, measurement of input excitation is not always possible. Identification methods that use only output measurements are thus preferred. In some studies, the aim is to identify the force but the system is known (Lu and Law 2006). Some other studies track changes in the structural parameters over time (Yang et al. 2004) or incorporate structural identification with a modification process (Chen and Li 2004). Most of them involve iterative least-squares procedures by making successive estimations of force and/or system parameters until achieving convergences.

Non-Classical Methods: Non-classical methods based on heuristic principles are promising and can be employed in structural system identification. The main characteristics of non-classical methods are the use of nature-inspired rules to solve high-dimensional problems which may otherwise be unattainable by classical methods. Relying heavily on computational resources, these methods make few or no assumptions in searching a large space for the global optimal solution of the identified system. Recent non-classical methods include simulated annealing (SA) method, particle swarm optimization (PSO) method, artificial neural networks (ANN) method, and genetic algorithms (GA). SA method was applied to structural damage identification as a global optimization technique (Bayissa and Haritos 2009). But the accuracy and efficiency of the damage severity estimation would be influenced by incomplete measurements and noise levels. PSO algorithm was applied to many structural and multidisciplinary optimization problems (Alrashidi and EI-Hawary 2006). But there are several explicit parameters impacting convergence of optimizing search and premature convergence is another problem. ANN method was usually applied to damage detection problems (Feng and Bahng 1999, Zubaydi et al. 2002). This method, however, is highly dependent on the training patterns and is therefore limited to the number of unknown parameters.

In particular, GA has been successfully applied to many optimization and search problems including structural identification. Based on Darwin's evolutionary theory of "survival of the fittest", this heuristic approach characterizes an intelligent exploitation of historical information to direct the search for better performance over generations involving mainly fitness-based selection, crossover (recombination of parental genes), and random mutation. Luh and Wu (1999) applied a GA-based scheme to identify the parameters of a non-linear autoregressive system. Hao and Xia (2002) identified damaged elements even with an imperfect analytical model by GA method. Perra and Torres (2006) applied GA to damage detection based on changes in frequencies and mode shapes. Nevertheless, for large structural systems with many unknown parameters, difficulty in convergence becomes a great challenge for most (if not all) identification methods including GA methods. To this end, Koh et al. (2000, 2003) developed GA-based substructural identification methods to reduce the size of the system to be identified. An improved substructural approach was then developed by using acceleration measurements to account for interactions between substructures without approximation of interface forces (Trinh and Koh 2012). In addition, it is beneficial to include a local search to improve convergence (Zhang et al. 2010). Recently, a new GA-based strategy called "search space reduction method" (SSRM) using an improved GA-based on migration and artificial selection (iGAMAS) was developed (Koh and Perry 2010). SSRM was shown to be an effective method by adaptively adjusting the search space to speed up the global optimum search.

1.2 Offshore Structural Systems

The offshore industry mainly involves the exploration and production of oil and gas in reservoirs below the sea floor. Classified as being either bottom-supported or floating, offshore structures are usually required to stay for a prolonged period in ocean environmental conditions including random waves, winds, and currents. Bottom-supported structures are generally fixed to the sea bed such as jackets and jack-ups, while floating structures are compliant by nature such as semi-submersible floating production and offloading unit (FPO), ship-shaped floating production, storage and offloading unit (FPSO), and spars and tension leg platforms. Bottom-supported and floating structures are very different not only in their appearance but also in their construction, installation, and dynamic characteristics. The common characteristics are that they all provide deck space and preload capacity to support equipment and variable weights for drilling and production operations.

Offshore structures are always affected by complex environmental loads. Static loads include gravity, hydrostatic and current loads, while dynamic loads arise from variable winds and waves. In practice, the current does not change rapidly with time and is usually treated as a constant or quasi-static load. It, therefore, has no influence on the dynamic response. Similarly, wind is predominately quasi-static and has little effect on the dynamic response. Since random wave is more dynamic, this research focuses on wave forces as the main source of external excitations over the contributions from winds and currents. Forward analysis is to predict dynamic response of offshore structures under time-varying environmental conditions. Inverse analysis of offshore structures involves system identification to determine unknown parameters based on measured structural response. System identification of offshore structures is very useful for structural health monitoring (SHM) which helps to detect, assess, and respond to potential dangers arising from structural damages due to environmental conditions or other causes. Jack-up rig, a self-installation platform, is a mobile drilling unit well suited for relatively shallow water (water depth/wave length < 0.05). Jack-up rig count has been making steady progress in the past few years mainly due to its better cost-effectiveness. Recently this type of rigs has been extended to use in deeper waters where harsher environment is expected. Safety assessment is thus very important for the continuing success of jack-up rigs. To this end, system identification of jack-up platform is beneficial to provide early identification of structural damage and hence reduce the risk of structural failure to an acceptable level.

Wave Forces on Offshore Structures: A number of wave theories have been developed for offshore structures (Dean and Dalrymple 1984, Chakrabarti 1987, Chakrabarti 2005). Three essential parameters are needed to describe a wave theory, i.e., wave period (T), wave height (H), and water depth (d). The simplest and most widely applied wave theory is linear wave theory, also called small amplitude wave theory or airy theory. Water is assumed to be incompressible, irrotational, and inviscous. A velocity potential and a stream function should exist for linear waves. Applying the necessary boundary conditions to the governing differential equation of water, many useful formulae can be derived for describing waves. For linear wave theory, the wave profile is formed by a sinusoidal function as

$$\eta = \frac{H}{2} \sin(kx - \omega t) \tag{1}$$

where $\omega = \frac{2\pi}{T}$ is circular frequency, L is wave length and $k = \frac{2\pi}{L}$ is wave number. Non-linear wave theories are also developed and applied to offshore structures such as second-order and fifth-order Stokes wave theories. As the names imply, these two wave theories comprise higher order components in a series form to describe the wave profile. However, higher order components are much smaller than the first order and decay rapidly with depth, so their effect in deeper water is negligible. Even when a non-linear wave theory is applied, higher order components have significant effect on the structure only near the free surface. Away from mean water level, wave behaves more like linear wave. When a single wave design approach is selected, regular wave can be used with single wave frequency, wave length, and wave height.

As for random ocean wave, it can be described by an energy density spectrum with statistical parameters. Several spectrum formulae are derived from the observed properties of ocean waves. Thus, they are empirical in nature. The widely used wave spectrum models include Pierson-Moskowitz (P-M) spectrum (Pierson and Moskowitz 1964), Bretschneider spectrum (Bretschneider 1959), ISSC spectrum, and JONSWAP spectrum (Hasselmann 1973). Different wave spectrum models are applied to different offshore locations when the site-specific wave spectrum is unavailable. For example, P-M spectrum is suitable for Gulf of Mexico, West Australia, and West Africa while JONSWAP spectrum is suitable for the North Sea and Northern North Sea. When dynamic analysis of offshore structure in the time domain is required, the time history of wave profile can be derived by wave theory and wave spectrum.

Wave loads are generally computed by two different methods depending on the size of the structure. Morison's equation (Morison et al. 1950) is an empirical formula to compute inertial and drag forces on small structures, and the commonly used expression is written as

$$df = (\rho C_m \nabla \dot{u} + \frac{1}{2} \rho C_d Au|u|)dl \tag{2}$$

where C_m is inertial coefficient, C_d is drag coefficient, A is projected area of wet structure per unit elevation, and ∇ is volume of wet structure per unit elevation. For large structures, as the structural dimension is very large compared to the wave length, the structure alters the form of the incident waves over a large area in its vicinity. So the wave forces have to be computed by diffraction/radiation theory. In this scenario, the potention flow should be used and derived by several numerical procedures such as boundary element method (BEM). The numerical procedures can be found in some references (Chakrabarti 1987, 2005).

Once the wave forces on the structure are known, structural response can be computed from the equation of motion for the dynamic system. The computation is generally categorized as deterministic analysis and stochastic analysis. Deterministic analysis is used to evaluate extreme conditions by considering individual wave heights and frequencies. It is necessary to design the offshore structure avoiding failures of construction and operation. To establish a more rational design procedure, stochastic analysis is a good alternative to statistically consider the irregular or random nature of wave forces. Wave field is assumed to be stationary in time domain, homogeneous in space and, distributed with Gaussian probability (Chakrabarti 1987). Stochastic analysis of offshore structures can be carried out in either time domain or frequency domain (Shinozuka et al. 1977, Barltrop and Adams 1991).

Offshore Jack-up Platform: Jack-up platform is a mobile self-elevating drilling unit used for offshore oil and gas exploration in shallow water. It typically comprises a buoyant hull supported by a number of vertically retractable truss-work legs. Jack-up rigs are classified as leg-independent jack-up and mat-supported jack-up according to different foundations. Independent leg jack-up platform is more popular due to easier installation and lower cost and is thus selected as a representative rig. It consists of three independent lattice legs with spudcan foundations and a triangular hull, as shown in Figure 1. The jack-up platform is transported to site floating on its hull with the legs elevated out of the water, and then positioned by lowering the legs onto the seabed and elevating the hull off the water. It is fixed in site by spudcan foundations. The spudcan penetrates into the seabed under self-weight of the jack-up with a preloading process of providing sufficient resistance capacity to reach the equilibrium.

The legs of jack-up platform are slender and more flexible than other fixed offshore structures. Dynamic effects become more important, since the natural period increases and may coincide with the wave period involving significant energy. With more accurate model parameters of leg flexibility, the performance and reliability of jack-up platform could be better assessed. Therefore, the flexibility of jack-up legs should be considered as a key parameter in system identification. The spudcan is usually considered as a pinned footing without rotational fixity to each leg (Senner 1993). However, it is a conservative assumption, as soil restraint may reduce the critical stresses at the leg-hull connection. Also, the increased fixity of spudcan fixity may increase the natural frequency of jack-up platform. More accurate estimation of spudcan fixity can improve dynamic analysis of jack-up

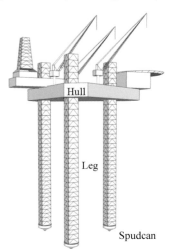

Figure 1. Schematic diagram of a typical jack-up unit.

platform to better predict structural response. In this regard, the fixity of spudcan foundation should be considered as another key parameter in system identification.

Offshore Structural System Identification: As the drilling operations of offshore structures are extended to deeper waters and harsher environments, structural health monitoring becomes increasingly important for safety considerations. System identification of offshore structures has the potential of non-destructive assessment to provide timely information on the structures and prevent any undesired accidents. Parameter identification of offshore structure was studied using Kalman filtering algorithms by Yun and Shinozuka (1980). Shyam Sunder and Sanni (1984) carried out foundation stiffness identification for fixed offshore structures based on ambient or force response measurements. But these studies required good initial guess and involved few unknown parameters. Mangal et al. (2001) attempted to identify the natural frequencies of offshore structures and compared them with the measured vibration signals to determine the damage locations. The difficulty in such an approach is that change in stiffness of individual member does not necessarily lead to a noticeable frequency change. Najafian (2007a,b) established a simple relationship between output and input of nonlinear offshore systems but the model was not verified by using any identification methods.

Due to mobility and better cost-effectiveness, offshore jack-up platforms have been widely used in shallow waters since around 1949 (Denton 1986). Three jack-up drilling units operating in the North Sea were monitored by permanently installed instruments (Springett et al. 1996, Temperton et al. 1999, Nelson et al. 2001). Using the case records from the three jack-up units at a total of eight locations, Cassidy et al. (2002) determined the appropriate stiffness levels for spudcan foundations by comparing the measured data and the numerical simulation results in both the frequency domain and the magnitude of response. Nataraja et al. (2004) calibrated the foundation fixity with adjusted soil spring stiffness resulting in the same measured and computed natural frequencies. Recently, experimental pushover loads and displacements on the hull and spudcans of a three-legged model jack-up were compared to numerical simulations with different stiffness assumptions of spudcan foundations on sand (Cassidy et al. 2010) and on clay (Vlahos et al. 2008, 2011). In these aforementioned studies, trial and error is often used to evaluate spudcan fixity by matching the measured natural frequencies and/or the measured magnitude of response with those computed by an analytical or numerical jack-up model. From an optimization point of view, however, this is not robust and effective in the global search for optimal solution, especially when involving many more unknowns.

In summary, there has been very little research on accurate parameter identification and damage detection of offshore structure systems in a global and quantitative way. The implementation of identification methods is highly recommended in the field of offshore engineering. It requires further research to develop robust and effective strategies for offshore structural system identification.

1.3 Objective and Scope

Based on the above literature review, it is noted that there is no known effective strategy for global system identification of offshore structures. In view of needs and benefits, it is important and necessary to develop robust strategies to identify key parameters and detect damage to offshore structures for the purpose of safety assessment and model updating. Due to the widespread and successful use of jack-up platform, it is selected to study in this research for identification of both leg stiffness and spudcan fixity as key parameters considering the challenging aspects. The main objective of this research is to develop robust and effective strategies for system identification of offshore jack-up platform.

To achieve the main objective, the scope of this research is as follows:

1. To perform dynamic analysis for a numerical model of jack-up platform as a basis of system identification.
2. To develop robust and effective strategies for parameter identification and damage detection of jack-up platform in both time domain and frequency domain. Leg stiffness parameters and spudcan fixities are the key parameters to identify.
3. To design a laboratory experimental study for validation of the proposed identification strategies.

2. Dynamic Analysis of Offshore Jack-up System

To carry out parameter identification and damage detection, dynamic analysis of a jack-up platform is first performed to analyze its behavior under different environmental and foundational conditions. The computed dynamic responses from these analyses are recorded as "measured" data in the numerical simulation study of structural identification. For this purpose, a finite element analysis of jack-up structural system is conducted. The equation of motion of the jack-up can be written as

$$\mathbf{M\ddot{u}} + \mathbf{C\dot{u}} + \mathbf{Ku} = \mathbf{F} \qquad (3)$$

where $\mathbf{M}$, $\mathbf{C}$ and $\mathbf{K}$ are the mass, damping, and stiffness matrices; $\mathbf{u}$ is the displacement vector; and $\mathbf{F}$ is the loading vector resulting from wind, wave, and current. In a jack-up analysis, not only the structure but also the environmental loading and foundational situation contribute to the dynamic Equation (3). The structural model, environmental loading, and spudcan foundation are briefly explained as follows.

2.1 Jack-up Structural Model

A three-legged jack-up platform, as shown in Figure 2, is considered in this study. Due to the dominating deformation of the slender lattice legs and the presence of significant axial forces in the jack-up, a beam-column stiffness formulation for large deformation analysis of three-dimensional elastic frames (Bienen and Cassidy 2006) is adopted to account for the P-Δ effect in jack-up legs. As shown in Figure 2, both the lattice legs and hull are modelled by elastic beam elements with equivalent characteristics (SNAME 2002).

The finite element method is applied to model the 3D frame structure using beam elements with six degrees of freedom (DOFs) at each node, i.e., three translations and three rotations. A consistent stiffness matrix for the beam element of uniform cross section is used in the model. The hull is represented by a number

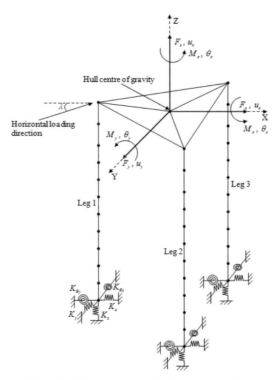

Figure 2. A 3D jack-up model with sign convention.

of beams provided with bending and torsional stiffnesses to correctly represent the "out of plane" stiffness of the hull, as illustrated in Figure 2. The leg masses are lumped at the nodes between the beam elements. The hull mass is divided over the centre node and the leg-to-hull connection nodes such that possible hull weight eccentricity and rotational inertial of the mass can be represented properly. Structural damping is modeled using a Rayleigh approach such that the damping matrix depends on the mass and stiffness matrices as

$$\mathbf{C} = \alpha\mathbf{M} + \beta\mathbf{K} \tag{4}$$

The above damping factors α and β can be related to the modal damping ratios (ξ_r) for any two modes (typically the first two modes $r = 1$ and 2) as

$$\xi_r = \frac{\alpha}{2\omega_r} + \frac{\beta\omega_r}{2} \tag{5}$$

Setting the damping for the two low modes renders the damping in higher modes artificially high (Clough and Penzien 1993). This is useful for jack-up dynamic analysis because the primary responses are in the surge and sway modes, which are the two lowest modes (Bienen and Cassidy 2006). Hence, the artificial damping helps to filter out inaccurate response of higher modes.

2.2 Offshore Environmental Loading

Environmental loads on offshore structures are caused by wind, current, and waves. Wind and current vary slowly and are often treated as static (or quasi-static) loads to the jack-up. On the other hand, waves are treated as dynamic loads and therefore the spectrum of wave loading must be modeled realistically. In this study, the sea state is simulated by a JONSWAP spectrum defined by significant wave height, peak period, and peak enhancement factor (Chakrabarti 2005). Linear wave theory is adopted to evaluate the surface elevations and wave kinematics.

When applying equivalent leg models to the jack-up, forces arising from vertical motion of wave particles are neglected as they usually tend to be less than 5% of the horizontal forces (Mommaas and Dedden 1989). A horizontal wave direction at an inclination λ (see Figure 2) is specified to investigate asymmetric loading conditions in the 3D jack-up model. The horizontal dimension of a jack-up leg is very small compared to the wave length and thus Morison equation (Morison et al. 1950) can be used to compute the hydrodynamic loads on the jack-up legs. The added mass and hydrodynamic damping matrices are also evaluated and incorporated into the respective system matrices. The structural displacement and velocity of jack-up platform are normally small and water particle kinematics can be evaluated at the undeformed position with the equivalent nodal loads applied along each leg in the finite element analysis. Quasi-static current load is superimposed to the wave load. Its distribution can be simulated as constant or linearly variable in depth. Based on an appropriate wind speed and using the projected area of the hull, equipment on the deck and exposed area of the legs (above mean water level), quasi-static wind load is applied at the hull.

2.3 Jack-up Spudcan Foundation

A jack-up platform is supported by spudcan foundations on the seabed. The characteristics of this spudcan to seabed connection are very important for the jack-up overall behaviour. When the foundation is vertically preloaded to a certain value, combinations of moments, horizontal, and vertical forces can be accommodated under elastic conditions as long as the soil stress level does not exceed the stress level reached during preloading (Mommaas and Dedden 1989). The soil stiffness level for an operating jack-up is one of the primary targets for identification. During a storm condition, an equivalent linearized value of the soil stiffness can also be identified.

The foundation fixity in three dimensions can be represented by five elastic springs, comprising three translational (K_x, K_y and K_z) and two rotational ($K_{\theta x}$ and $K_{\theta y}$) as shown in Figure 2, for soil-structure interaction. For dynamic analysis, the foundation spring stiffness is evaluated in accordance with Poulos and Davis (1974) and SNAME (2002):

$$
\begin{Bmatrix} K_x \\ K_y \\ K_z \\ K_{\theta x} \\ K_{\theta y} \end{Bmatrix} = \begin{Bmatrix} \dfrac{16 G_x B(1-\upsilon)}{(7-8\upsilon)} \\[2mm] \dfrac{16 G_y B(1-\upsilon)}{(7-8\upsilon)} \\[2mm] \dfrac{2 G_z B}{(1-\upsilon)} \\[2mm] \dfrac{G_{\theta x} B^3}{3(1-\upsilon)} \\[2mm] \dfrac{G_{\theta y} B^3}{3(1-\upsilon)} \end{Bmatrix} \tag{6}
$$

where B is the effective spudcan diameter at the uppermost part of bearing area, υ is the Poisson's ratio of soil, G_x and G_y are the shear moduli for horizontal loadings in x- and y-direction, respectively, G_z is the shear modulus for vertical (z-direction) loading, and $G_{\theta x}$ and $G_{\theta y}$ are the shear moduli for rotational loadings.

When the spacing between jack-up legs is large in relation to lateral dimensions of spudcan, the contribution of torsional stiffness of individual spudcan foundation in resisting the global torsion of the jack-up is small (Karthigeyan 2009). Therefore, the torsional stiffness ($K_{\theta z}$) of spudcan foundation is not presented in the Equation (6). The soil spring stiffness computed using the above equation is incorporated into the system stiffness matrix. The spudcan dynamic mass is lumped into the leg-spudcan connection point in the structural model. Note that in this study, the different amount of spudcan stiffness with different G moduli can be identified based on the measured jack-up responses. The nominal amount of spudcan stiffness is first calculated using Equation (6) with different G moduli. They are only used as the initial guess for the identification procedure. Finally, the optimal solution to the spudcan stiffness (and thus G modulus) is identified based on a broad search range of half to two times of the nominal value.

2.4 Dynamic Analysis in Time Domain

For dynamic analysis of large scale structures, an effective way is to divide the large structural system into some smaller structural systems (Koh and Perry 2010). As an example, a simple linear structure is shown in Figure 3(a) and the substructure is extracted in Figure 3 (b). The equation of motion (3) can be written as

$$
\begin{bmatrix}
\mathbf{M}_{uu} & \mathbf{M}_{uf} & & & \\
\mathbf{M}_{fu} & \mathbf{M}_{ff} & \mathbf{M}_{fr} & & \\
& \mathbf{M}_{rf} & \mathbf{M}_{rr} & \mathbf{M}_{rg} & \\
& & \mathbf{M}_{gr} & \mathbf{M}_{gg} & \mathbf{M}_{gl} \\
& & & \mathbf{M}_{lg} & \mathbf{M}_{ll}
\end{bmatrix}
\begin{Bmatrix}
\ddot{u}_u \\ \ddot{u}_f \\ \ddot{u}_r \\ \ddot{u}_g \\ \ddot{u}_l
\end{Bmatrix}
+
\begin{bmatrix}
\mathbf{C}_{uu} & \mathbf{C}_{uf} & & & \\
\mathbf{C}_{fu} & \mathbf{C}_{ff} & \mathbf{C}_{fr} & & \\
& \mathbf{C}_{rf} & \mathbf{C}_{rr} & \mathbf{C}_{rg} & \\
& & \mathbf{C}_{gr} & \mathbf{C}_{gg} & \mathbf{C}_{gl} \\
& & & \mathbf{C}_{lg} & \mathbf{C}_{ll}
\end{bmatrix}
\begin{Bmatrix}
\dot{u}_u \\ \dot{u}_f \\ \dot{u}_r \\ \dot{u}_g \\ \dot{u}_l
\end{Bmatrix}
$$

$$
+
\begin{bmatrix}
\mathbf{K}_{uu} & \mathbf{K}_{uf} & & & \\
\mathbf{K}_{fu} & \mathbf{K}_{ff} & \mathbf{K}_{fr} & & \\
& \mathbf{K}_{rf} & \mathbf{K}_{rr} & \mathbf{K}_{rg} & \\
& & \mathbf{K}_{gr} & \mathbf{K}_{gg} & \mathbf{K}_{gl} \\
& & & \mathbf{K}_{lg} & \mathbf{K}_{ll}
\end{bmatrix}
\begin{Bmatrix}
u_u \\ u_f \\ u_r \\ u_g \\ u_l
\end{Bmatrix}
=
\begin{Bmatrix}
P_u \\ P_f \\ P_r \\ P_g \\ P_l
\end{Bmatrix}
\tag{7}
$$

where r denotes internal DOFs in substructure, f, g denotes interface DOFs in substructure, and u, l denote non-substructure part. The equations of motion for substructure can be extracted from Equation (7) as follows:

$$\begin{bmatrix} \mathbf{M}_{ff} & \mathbf{M}_{fr} & \\ \mathbf{M}_{rf} & \mathbf{M}_{rr} & \mathbf{M}_{rg} \\ & \mathbf{M}_{gr} & \mathbf{M}_{gg} \end{bmatrix} \begin{Bmatrix} \ddot{\mathbf{u}}_f \\ \ddot{\mathbf{u}}_r \\ \ddot{\mathbf{u}}_g \end{Bmatrix} + \begin{bmatrix} \mathbf{C}_{ff} & \mathbf{C}_{fr} & \\ \mathbf{C}_{rf} & \mathbf{C}_{rr} & \mathbf{C}_{rg} \\ & \mathbf{C}_{gr} & \mathbf{C}_{gg} \end{bmatrix} \begin{Bmatrix} \dot{\mathbf{u}}_f \\ \dot{\mathbf{u}}_r \\ \dot{\mathbf{u}}_g \end{Bmatrix}$$

$$+ \begin{bmatrix} \mathbf{K}_{ff} & \mathbf{K}_{fr} & \\ \mathbf{K}_{rf} & \mathbf{K}_{rr} & \mathbf{K}_{rg} \\ & \mathbf{K}_{gr} & \mathbf{K}_{gg} \end{bmatrix} \begin{Bmatrix} \mathbf{u}_f \\ \mathbf{u}_r \\ \mathbf{u}_g \end{Bmatrix} = \begin{Bmatrix} \mathbf{P}_f \\ \mathbf{P}_r \\ \mathbf{P}_g \end{Bmatrix} \tag{8}$$

Let j denote all interface DOFs including both f and g. The second row of Equation (8) can be further rearranged by taking the interface effects as additional input excitation for the internal nodes

$$\mathbf{M}_{rr}\ddot{\mathbf{u}}_r(t) + \mathbf{C}_{rr}\dot{\mathbf{u}}_r(t) + \mathbf{K}_{rr}\mathbf{u}_r(t) = \mathbf{P}_r(t) - \mathbf{M}_{rj}\ddot{\mathbf{u}}_j(t) - \mathbf{C}_{rj}\dot{\mathbf{u}}_j(t) - \mathbf{K}_{rj}\mathbf{u}_j(t) \tag{9}$$

If the time history of interface acceleration ($\ddot{u}_j(t)$) is given, the velocity and displacement can be evaluated by trapezoidal rule (Trinh and Koh 2011) as

$$\dot{u}_j^{k+1} = \dot{u}_j^k + \frac{\Delta t}{2}\left(\ddot{u}_j^k + \ddot{u}_j^{k+1}\right)$$

$$u_j^{k+1} = u_j^k + \frac{\Delta t}{2}\left(\dot{u}_j^k + \dot{u}_j^{k+1}\right) \tag{10}$$

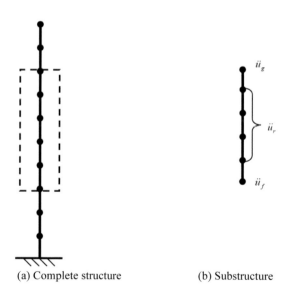

(a) Complete structure   (b) Substructure

Figure 3. Substructural analysis.

where superscripts k and $k + 1$ represent two consecutive data points, and Δt is time step. Therefore, accelerations, velocities, and displacements at the interface are all available and can be substituted in Equation (9) as the interface force. The substructural response can be computed by applying Newmark method to the equation of motion. If the measurements of interface accelerations are clean signals, the computed results should be the same as the complete structure method. Practical measurements are inevitably contaminated by noises, thus the numerical integration will introduce drift in velocity and displacement time histories, leading to a low-frequency drift in the interface force vector. Nevertheless, the natural frequencies of a substructure are higher than those of the complete structure, and substructural response is predominantly excited by force components with frequencies close to the substructure's natural frequencies. Hence, the low-frequency drift in integrated force vector does not have any significant effects on substructural dynamic response.

2.5 Dynamic Analysis in Frequency Domain

For spectral analysis (Clough and Penzien 1993), the equation of motion (3) can be transformed into frequency domain with Fourier transform of structural acceleration as the main variable:

$$\left(\mathbf{M} + \frac{\mathbf{C}}{i\omega} - \frac{\mathbf{K}}{\omega^2} \right) \hat{\ddot{\mathbf{u}}}(i\omega) = \hat{\mathbf{P}}(i\omega) \tag{11}$$

where $\hat{\ddot{\mathbf{u}}}(i\omega)$ is the Fourier transform of structural acceleration, $\hat{\mathbf{P}}(i\omega)$ is the Fourier transform of input excitation, and ω is circular frequency (rad/s). As a result, structural response vector in frequency domain can be derived by

$$\hat{\ddot{\mathbf{u}}}(i\omega) = \mathbf{G}(i\omega)\,\hat{\mathbf{P}}(i\omega) \tag{12}$$

where $\mathbf{G}(i\omega)$ is frequency response transfer matrix. Therefore, for spectral analysis, spectral density matrix of structural response can be obtained by

$$\mathbf{S}_{\ddot{u}}(i\omega) = \mathbf{G}(i\omega)\mathbf{S}_{p}(i\omega)\mathbf{G}^{T}(-i\omega) \tag{13}$$

where $\mathbf{S}_{p}(i\omega)$ and $\mathbf{S}_{\ddot{u}}(i\omega)$ are the power spectral density matrices of input excitation and structural response respectively:

$$
\mathbf{S_p}(i\omega) = \begin{bmatrix} S_{P_1 P_1}(\omega) & S_{P_1 P_2}(i\omega) & \cdots & S_{P_1 P_N}(i\omega) \\ S_{P_2 P_1}(i\omega) & S_{P_2 P_2}(\omega) & \cdots & S_{P_2 P_N}(i\omega) \\ \cdots & \cdots & \cdots & \cdots \\ S_{P_N P_1}(i\omega) & S_{P_N P_2}(i\omega) & \cdots & S_{P_N P_N}(\omega) \end{bmatrix} \tag{14}
$$

where

$$
S_{P_j P_k}(i\omega) = \lim_{s \to \infty} \frac{\left[\int_{-s/2}^{s/2} P_j(t) \exp(-i\omega t)\,dt \right]\left[\int_{-s/2}^{s/2} P_k(t) \exp(i\omega t)\,dt \right]}{2\pi s}
$$

$$
\mathbf{S_{\ddot{u}}}(i\omega) = \begin{bmatrix} S_{\ddot{u}_1 \ddot{u}_1}(\omega) & S_{\ddot{u}_1 \ddot{u}_2}(i\omega) & \cdots & S_{\ddot{u}_1 \ddot{u}_N}(i\omega) \\ S_{\ddot{u}_2 \ddot{u}_1}(i\omega) & S_{\ddot{u}_2 \ddot{u}_2}(\omega) & \cdots & S_{\ddot{u}_2 \ddot{u}_N}(i\omega) \\ \cdots & \cdots & \cdots & \cdots \\ S_{\ddot{u}_N \ddot{u}_1}(i\omega) & S_{\ddot{u}_N \ddot{u}_2}(i\omega) & \cdots & S_{\ddot{u}_N \ddot{u}_N}(\omega) \end{bmatrix} \tag{15}
$$

According to drag and inertia terms used in time domain analysis, spectral density matrix for waveforce can be obtained directly from wave spectrum as follows:

$$
S_{P_j P_k}(i\omega_i) = \left(C_d \frac{\rho A}{2} \sqrt{\frac{8}{\pi}}\sigma_U \right)_j \left(C_d \frac{\rho A}{2} \sqrt{\frac{8}{\pi}}\sigma_U \right)_k S_{U_j U_k}(i\omega_i) + (C_m \rho V)_j (C_m \rho V)_k S_{\dot{U}_j \dot{U}_k}(i\omega_i)
$$

$$
+ \left(C_d \frac{\rho A}{2} \sqrt{\frac{8}{\pi}}\sigma_U \right)_j (C_m \rho V)_k S_{U_j \dot{U}_k}(i\omega_i) + (C_m \rho V)_j \left(C_d \frac{\rho A}{2} \sqrt{\frac{8}{\pi}}\sigma_U \right)_k S_{\dot{U}_j U_k}(i\omega_i) \tag{16}
$$

where subscripts j, k denote node numbers of beam elements, and σ_{U_j} is the standard deviation of U_j,

$$
S_{U_j U_k}(i\omega_i) = S_{\eta\eta}(\omega_i) \left\{ \left[\omega_i \frac{\cosh(\kappa_i(z_j + d))}{\sinh(\kappa_i d)} \right]\left[\omega_i \frac{\cosh(\kappa_i(z_k + d))}{\sinh(\kappa_i d)} \right] \right\} e^{[i\kappa_i(x_j - x_k)]},
$$

$$
\sigma_{U_j} = \left[\sum_1^{N_f} S_{U_j U_j} \Delta\omega \right]^{1/2}, \; S_{U_j \dot{U}_k}(i\omega_i) = i\omega_i S_{U_j U_k}(i\omega_i), \; S_{\dot{U}_j U_k}(i\omega_i) = -i\omega_i S_{U_j U_k}(i\omega_i),
$$

$$
S_{\dot{U}_j \dot{U}_k}(i\omega_i) = \omega_i^2 S_{U_j U_k}(i\omega_i)
$$

For substructural analysis, Equation (9) can be directly transformed into frequency domain as

$$\left(\mathbf{M}_{rr} + \frac{\mathbf{C}_{rr}}{i\omega} - \frac{\mathbf{K}_{rr}}{\omega^2}\right)\hat{\ddot{\mathbf{u}}}_r(i\omega) = \hat{\mathbf{P}}_r(i\omega) - \left(\mathbf{M}_{rj} + \frac{\mathbf{C}_{rj}}{i\omega} - \frac{\mathbf{K}_{rj}}{\omega^2}\right)\hat{\ddot{\mathbf{u}}}_j(i\omega) \tag{17}$$

where $\hat{\ddot{\mathbf{u}}}_r(i\omega)$ is the Fourier transform of substructural acceleration, $\hat{\ddot{\mathbf{u}}}_r(i\omega)$ is the Fourier transform of interface acceleration, and $\hat{\mathbf{P}}(i\omega)$ is the Fourier transform of wave force inside substructure. Substructural response vector in frequency domain can be derived by

$$\hat{\ddot{\mathbf{u}}}_r(i\omega) = \mathbf{G}_r(i\omega)\hat{\mathbf{P}}_r(i\omega) - \mathbf{G}_j(i\omega)\hat{\ddot{\mathbf{u}}}_r(i\omega) \tag{18}$$

where $\mathbf{G}_r(i\omega)$ is frequency response matrix for internal nodes, and $\mathbf{G}_j(i\omega)$ is frequency response matrix for interface nodes. The power spectral density matrix of substructural response can be obtained by

$$\mathbf{S}_{\ddot{\mathbf{u}}_r}(i\omega) = \mathbf{G}_r(i\omega)\mathbf{S}_{\mathbf{p}_r}(i\omega)\mathbf{G}_r(-i\omega)^T - \mathbf{G}_j(i\omega)\mathbf{S}_{\ddot{\mathbf{u}}_j}(i\omega)\mathbf{G}_j(-i\omega)^T \tag{19}$$

where $\mathbf{S}_{\mathbf{p}_r}(i\omega)$ and $\mathbf{S}_{\ddot{\mathbf{u}}_j}(i\omega)$ are derived from Equations (14) and (15) respectively.

3. Parameter Identification and Damage Detection of Offshore Jack-up System: Time Domain

Due to the large size of jack-up platforms, substructural identification (Koh et al. 1991) is adopted to improve the accuracy and efficiency of identification significantly by reducing the system size. This approach is a divide-and-conquer strategy to identify critical parts rather than attempting the very difficult (if not impossible) task of identifying the whole structure. Furthermore, this approach has the advantage of avoiding some complex boundary conditions (which are difficult to model accurately) such as leg-hull connections by excluding them from the substructures. A single leg in 2D jack-up model as shown in Figure 4(a) is divided into two substructures. As illustrated in Figure 4(b), the top half of the leg is Substructure 1 (S1) and the bottom half of the leg with spudcan fixity is Substructure 2 (S2). It should be noted that 2D frame jack-up model is, in fact, another form of substructuring which makes the analysis and identification much more efficient than the 3D jack-up model (see Figure 2). The plane (or direction) of consideration for the selected leg is chosen according to the measurements in the corresponding plane. Therefore, the 3D model analysis can be avoided for identification of a single leg (or part of it) in the plane of consideration. Where parameters in another plane are to be identified, the same procedure is repeated in the new plane. Moreover, besides the key unknown parameters (i.e., leg stiffness

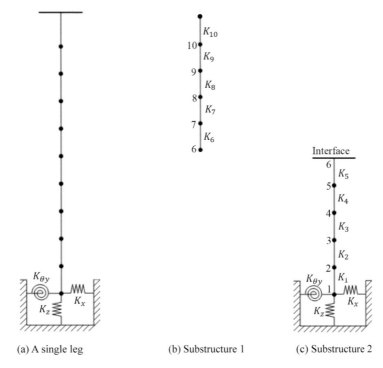

(a) A single leg   (b) Substructure 1   (c) Substructure 2

Figure 4. 2D Substructures (in XZ plane) of jack-up model.

and spudcan fixity), damping coefficients (α, β) are treated as unknown in the identification procedure.

3.1 Substructural Identification Strategy in Time Domain

The proposed strategy should address several challenges unique to offshore structures (as opposed to land-based structures). Structural identification in time domain usually requires known initial conditions. This requirement is, however, not realistic for offshore structures subjected to wave loading continually. A procedure is proposed herein to deal with unknown initial conditions. Besides, in practice, wave forces are difficult to measure or to accurately predict. Thus, output-only strategy is preferred. A predictor-corrector algorithm (Perry and Koh 2008) is adopted herein to tackle this problem in time domain.

Changes in structural stiffness values influence structural dynamic response. The sum of squared errors (SSE) between the measured and simulated time histories of structural response is a good indicator to guide the GA-based search for unknown parameter values. Besides, the normalized structural response is preferred to address the problems arising from possible differences in order of magnitudes at different nodes or different units for different response quantities (e.g., linear acceleration versus angular acceleration). Since the dynamic response is normally measured by accelerometers, accelerations are adopted in fitness definition. The following fitness function is used in time domain identification:

$$Fitness = \frac{1}{c + \sum_{i=1}^{N} \sum_{j=1}^{N_T} \frac{\left(\ddot{u}_{i,j}^{m} - \ddot{u}_{i,j}^{s}\right)^2}{E\left(\ddot{u}^2\right)_i^m}} \tag{20}$$

where $E\left(\ddot{u}^2\right)_i^m = \sum_{k=1}^{T}\left(\ddot{u}_{i,k}^{m}\right)^2 \Big/ N_T$ represents the mean squared value of measured accelerations, N is the number of measured responses, N_T is the data length, superscripts m and s represent the measured and simulated values respectively, and c is an arbitrary constant to avoid singularity in the above equation and is chosen to be of the same order of magnitude as squared value of normalized accelerations.

The output-only strategy involving a predictor-corrector algorithm is adopted. Based on measured accelerations, interface velocities and displacements can be obtained by the trapezoidal rule of integration as in Equation (10). Similarly, an initial estimate of the velocities and displacements at internal nodes at time step $k+1$ is first predicted from the measured accelerations at time step $k+1$ and "corrected" response (as it will be explained later) at time step k by using the trapezoidal rule of integration. Then, the unknown force at time step $k+1$ is predicted by

$$\mathbf{P}_r = \mathbf{M}_{rr}\ddot{\mathbf{u}}_r + \mathbf{C}_{rr}\dot{\mathbf{u}}_r + \mathbf{K}_{rr}\mathbf{u}_r + \mathbf{M}_{rj}\ddot{\mathbf{u}}_j + \mathbf{C}_{rj}\dot{\mathbf{u}}_j + \mathbf{K}_{rj}\mathbf{u}_j \tag{21}$$

Assuming the force locations are known, the force vector should be corrected by putting zeros in locations without forces. Based on the corrected force and total interface response, the "corrected" response at internal nodes at this time step can be recomputed by integrating Equation (9) by Newmark's constant-average acceleration method. The corrected response is then passed on to the next time step and the process is repeated for the entire time history. As a result, the corrected accelerations are stored to be compared with the measured accelerations in the fitness function for GA-based search.

It should be noted that the wave spectrum required in the beginning of identification is estimated by some empirical data (such as an estimate of wave height for JONSWAP spectrum) to account for hydrodynamic damping. The precise information of wave spectrum is not needed, as the hydrodynamic damping term also includes the drag coefficient C_d which is one of the unknown parameters for identification. The numerical study shows the identification accuracy of C_d, and hence the accuracy of the estimated wave spectrum does not influence the stiffness identification significantly. A filter window is applied before the fitness computation to obtain a structural response in a specified frequency range covering selected natural frequencies of the substructures. Though initial conditions are unknown, the jack-up model is assumed to start from rest in the dynamic analysis required in the GA-based identification procedure. The transient state due to initial conditions would vanish due to damping effects. Hence, the problem of unknown initial conditions can be overcome by not including the transient state (initial response) in the fitness computation. Moreover, SSRM (Koh and Perry 2010) as mentioned earlier is adopted to improve the efficiency of the required GA-based search. The substructural identification procedure in time domain is summarized in Figure 5.

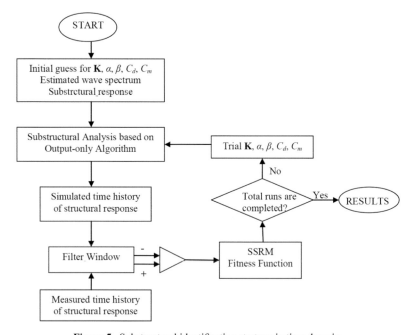

Figure 5. Substructural identification strategy in time domain.

3.2 Parameter Identification of Substructures

As an example, Magellan jack-up operation at Franklin site in the North Sea is studied. The geotechnical data and environmental data are from some previous research (Cassidy et al. 2002, Nataraja et al. 2004, Zhang et al. 2012, Wang et al. 2014). For the jack-up model as shown in Figure 2, structural properties used in this research are also summarized in Table 1. The wave is assumed to be governed

Table 1. Structural properties of jack-up model.

Description	Symbol	Value
For hull:		
Second moment of area	I_h	150 m⁴
Cross-sectional area	A_h	6 m²
Length	L_h	42 m
Mass	M_h	13650 t
For each leg:		
Young's modulus	E	200 Gpa
Second moment of area	I_l	15 m⁴
Cross-sectional area	A_l	0.6 m²
Length	L_l	120 m
Mass	M_l	1200 t
Equivalent diameter	D_e	8.50 m
Equivalent area	A_e	3.66 m²
For each spudcan:		
Horizontal stiffness	K_x	3.24×10^8 N/m
Vertical Stiffness	K_y	6.97×10^8 N/m
Rotational stiffness	K_r	2.87×10^{10} Nm/rad
Mass with enclosed water	M_s	600 t

by JONSWAP spectrum applicable to the North Sea. The water depth considered is 92 m. A storm condition is considered with significant wave height 7.59 m, peak wave period 11.13 s, and peak enhancement factor 4.7. Structural stiffness matrix and mass matrix (including water added mass) are formed. The damping factor including structure and foundation is assumed to be 3% (SNAME 2002). Hydrodynamic damping is also added to the nodes under water. The dynamic response can be derived by a complete structural analysis and further can be used as "measurement" in structural identification.

In practice, the measurements are inevitably contaminated by noise. In order to account for the noise effects in the numerical simulation study, a white Gaussian noise vector is applied and freshly generated in every numerical test, so as to avoid using the same pattern of noise. The noise is defined by

$$Noise = w \times RMS_{CleanSignal} \times NoiseLevel \qquad (22)$$

Where *RMS* denotes root mean square and *w* is a standard random Gaussian variable.

To define the fitness function required in GA-based substructural identification strategy, it is useful to study the sensitivity of structural response with respect to a change in stiffness (Wang 2012). The sensitivity study shows that rotational acceleration is more sensitive to leg stiffness change in S1 than linear accelerations. This means that, for S1 identification, rotational accelerations of internal nodes are preferably included in the fitness function. The sensitivity study for S2, which involves spudcan stiffness, shows that rotational response and horizontal response are more sensitive to leg stiffness change, while horizontal response and vertical response are more sensitive to spudcan stiffness change. A set of such response combination is, therefore, selected for S2 identification, i.e., horizontal accelerations of internal nodes excluding the first node are used in the fitness function with additional two pairs of vertical and rotational accelerations at the third node and fifth node.

Substructural identification is carried out with the parameters used in SSRM (Koh and Perry 2010) listed in Table 2. The search limits for unknown parameters are defined from half to double of the exact values. Two different noise levels of 5% and 10% are tested, and ten identification tests are conducted for each noise level. The identified stiffness values are different for different tests because the initial guess and noise sequences are both randomly generated in each test. Moreover, in view of the stochastic nature of GA-based identification, it is beneficial to average the identified parameters from different tests to achieve better identification results. The identification results of the ten tests are derived and presented in terms of ratios between identified values of mean parameters and exact values as shown in Figure 6. The identification results are very close to one, which means that structural stiffness can be accurately identified. The identification errors of mean parameters are also presented in Table 3. The errors of mean parameters based on time domain identification are less than 8% for 10% noise level and less than 4% for 5% noise level.

Table 2. GA Parameters Used in SSRM.

Total Runs	15
Runs	5
Generations	100
Populations	3×10
Regeneration Number	2
Reintroduction Number	30
Migration Rate	0.05
Crossover Rate	0.8
Mutation Rate	0.2
Window	4.0

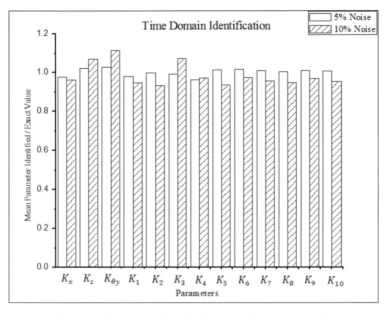

Figure 6. Time domain identification results in numerical study.

Table 3. Errors of mean parameters (%) in time domain.

Noise Level	K_x	K_s	$K_{\theta x}$	K_1	K_2	K_3	K_4	K_5	K_6	K_7	K_8	K_9	K_{10}
5%	2.37	2.08	2.80	1.93	0.20	0.85	3.74	1.28	1.62	1.05	0.40	1.07	0.54
10%	3.22	6.35	5.99	5.43	6.89	7.08	3.01	6.64	2.82	4.57	5.38	3.29	4.63

3.3 Damage Detection in Time Domain

Based on accurate parameter identification, damage assessment can be computed by comparing the identified parameters between the damaged structure and the undamaged structure. In this respect, damage is defined as the loss in structural stiffness as a percentage of the undamaged structure, represented by

$$D = \frac{K_d - K_u}{K_u} \tag{23}$$

where subscript d and u denote the damaged structure and undamaged structure respectively. The damage detection strategy is based on available measurements before and after damage. The structure mass is assumed to be known and unchanged. The main objective is to identify the stiffness changes between the damaged structure and the undamaged structure. Different from identification of the undamaged structure, identification of the damaged structure is carried out by including the identified parameters of the undamaged structure in the initial population. The search limits for unknown parameters are also set as half to double of the actual parameters of the undamaged structure.

Element K_8 near the water surface is considered to be the damage location in S1. The spudcan fixity is important for the overturning resistance of jack-up platform, especially the rotational stiffness. Thus, the damage considered in S2 is the loss of rotational stiffness of the spudcan fixity $(K_\theta)_y$. Damage is simulated as 20% loss of structural stiffness and two cases are studied. I/O noise is set at 5%, and the identified parameters of undamaged structure at 5% noise level are used in damage detection. The measurements based on the damaged structure are applied in the fitness computation. In total, 10 tests of damage detection are carried out in time domain for each damage case. The damage detection results in time domain are derived and listed in Table 4.

In Case 1, the damage is well detected in S1 through significant stiffness loss of K_8 and highlighted in bold in Table 4(a). The largest difference between the identified damage and exact damage of K_8 is less than 3%, and the mean damage is –19.8%, very close to the exact damage of 20%. Similarly for Case 2,

Table 4. Damage detection results in time domain.

(a) Case 1: identified damage (%) in S1

	1	2	3	4	5	6	7	8	9	10	Mean
K_6	–4.17	–3.50	–5.90	–4.84	–3.82	–3.23	–2.41	–0.15	–0.09	0.11	–2.82
K_7	–3.68	–0.37	–4.43	–1.89	1.99	–0.91	0.01	3.73	0.70	3.32	–0.15
K_8	**–21.83**	**–20.66**	**–21.43**	**20.61**	**17.70**	**19.87**	**20.53**	**18.37**	**–19.89**	**–17.12**	**–19.80**
K_9	–1.38	0.02	–1.95	0.54	2.15	–0.94	0.03	3.15	0.01	4.28	0.59
K_{10}	–4.32	0.55	–1.54	0.00	4.20	–0.20	1.04	1.45	–0.32	2.73	0.36

(b) Case 2: identified damage (%) in S2

	1	2	3	4	5	6	7	8	9	10	Mean
K_1	2.74	4.29	4.49	0.45	2.53	–1.73	1.22	2.54	2.00	1.67	2.02
K_2	0.78	6.43	5.10	–1.67	5.40	4.75	0.01	3.36	6.74	–1.69	2.92
K_3	–2.04	–7.94	–7.19	6.27	–1.63	–3.78	5.12	–5.27	2.03	–3.04	–1.75
K_4	4.69	0.02	–1.16	1.97	–4.80	0.08	0.03	–4.32	7.22	–4.24	–0.05
K_5	–5.06	1.24	–4.54	0.99	–0.91	1.47	–1.78	3.13	–4.19	–1.25	–1.09
K_x	5.50	5.19	–3.65	7.28	6.73	–4.82	–6.83	–6.50	4.36	–6.04	0.12
K_z	2.17	0.02	1.27	0.37	2.18	1.88	0.03	6.16	–5.20	2.35	1.12
$K_{\theta y}$	**–22.50**	**–22.88**	**–17.11**	**–18.56**	**–19.93**	**–24.38**	**–23.36**	**–21.71**	**–17.36**	**–21.35**	**–20.92**

the damage of spudcan as the stiffness loss of is also well-detected. The largest difference between the identified damage and exact damage in S2 is less than 5%, while the mean detected damage is –20.92%. Although some false damages are also identified as shown in Table 4, the largest value of mean false damages is less than 3%, which is less than the 5% I/O noise level. Moreover, it is noted that the identified damage values for false damage locations vary as positive or negative for different tests. But for real damage locations, the identified damage values are always negative, consistently pointing to the occurrence of damage. This means that comparison of identified damage values obtained by repeating the damage identification using different measurement signals is a useful check. Hence, the proposed identification strategy is a useful non-destructive way of damage detection for the existing offshore platforms.

4. Parameter Identification and Damage Detection of Offshore Jack-up System: Frequency Domain

As an alternative, spectral analysis can be used to solve the dynamic system with unknown initial conditions and random excitations. But there is no known procedure thus far to address unknown wave loading in frequency domain. This problem can be overcome by excluding unknown wave loading in frequency domain identification. Moreover, the natural frequencies of the substructures are much higher than those of the complete structure. The frequency range considered in system identification is selected to cover the main natural frequencies of the substructures. The dynamic response is, therefore, mainly excited by the higher-frequency excitations (e.g., mechanical systems and crane activities) on the jack-up hull and, if necessary, an additional force may be beneficially applied on the hull to generate a greater dynamic response for the substructures. Hydrodynamic effects (drag and added mass of water) are treated as unknown by including hydrodynamic coefficients (C_d, C_m) as unknown in the identification procedure.

4.1 Substructural Identification Strategy in Frequency Domain

Equation (9) of substructural analysis can be transformed into frequency domain as follows

$$\left(\mathbf{M}_{rr} + \frac{\mathbf{C}_{rr}}{i\omega} - \frac{\mathbf{K}_{rr}}{\omega^2} \right) \hat{\mathbf{u}}_r (i\omega) = \hat{\mathbf{P}}_r (i\omega) - \left(\mathbf{M}_{rj} + \frac{\mathbf{C}_{rj}}{i\omega} - \frac{\mathbf{K}_{rj}}{\omega^2} \right) \hat{\mathbf{u}}_j (i\omega) \tag{24}$$

Substructural response vector in frequency domain can be derived by

$$\hat{\mathbf{u}}_r (i\omega) = \mathbf{G}_r (i\omega) \hat{\mathbf{P}}_r (i\omega) - \mathbf{G}_j (i\omega) \hat{\mathbf{u}}_j (i\omega) \tag{25}$$

where $\mathbf{G}_r(i\omega)$ is frequency response matrix for internal nodes, and $\mathbf{G}_j(i\omega)$ is frequency response matrix for interface nodes. Therefore, spectral density matrix of substructural response can be obtained by

$$\mathbf{S}_{\ddot{u}_r}(i\omega) = \mathbf{G}_r(i\omega)\mathbf{S}_{\mathbf{p}_r}(i\omega)\mathbf{G}_r(-i\omega)^T - \mathbf{G}_j(i\omega)\mathbf{S}_{\ddot{u}_j}(i\omega)\mathbf{G}_j(-i\omega)^T \tag{26}$$

In frequency domain, it is not necessary to derive total interface response and only acceleration measurements are required. Note that the natural frequencies of the substructures are much higher than those of the complete structure. The frequency range is chosen to cover selected natural frequencies of the substructures but not the dominant wave frequency. In this way, wave force measurement is not required.

In frequency domain identification, the response measurements need to be transformed into power spectral density (PSD) values. Independent records of time history are converted by FFT to their PSD records. No averaging of these independent spectral records is needed, and no wave force measurement is required as mentioned earlier. Hence, the analysis can be used for non-stationary condition. The fitness function involves comparisons between measured and simulated PSD values of accelerations in a specified frequency range, which is defined by

$$Fitness = \cfrac{1}{c + \sum\limits_{i=1}^{N_f}\sum\limits_{j=1}^{N}\sum\limits_{k=j}^{N}\left(\cfrac{\left|S_{\ddot{u}_j\ddot{u}_k}^m(\omega_i)\Delta\omega - S_{\ddot{u}_j\ddot{u}_k}^s(\omega_i)\Delta\omega\right|}{E\left(\ddot{u}_j\ddot{u}_k\right)^m}\right)} \tag{27}$$

where $E\left(\ddot{u}_j\ddot{u}_k\right)^m = \sum\limits_{l=1}^{N_f}S_{\ddot{u}_j\ddot{u}_k}^m(\omega_l)\Delta\omega$ is the mean squared value of measured accelerations, $\Delta\omega$ is the frequency interval, N_f is the number of frequency points, N is the number of selected response points, and superscripts m and s represent measured and simulated PSD values respectively. A constant c is introduced to avoid possible zero denominator (in the ideal case when the measured and simulated PSD values are identical) and is chosen to be of the same order of magnitude as the squared value of normalized accelerations.

Note that the fitness function includes not only auto-spectral density but also cross-spectral density between different measurements. At each frequency point, the simulated spectral density matrix is compared with the measured spectral density matrix for selected responses of internal nodes in the fitness function. Since spectral density matrix of structural response is symmetric, only the upper triangular part is compared in the fitness function. Similar to the time domain strategy, the hydrodynamic damping is based on an estimate of wave spectrum and an unknown drag coefficient to be identified. Besides, Parzen window (Parzen 1962, Priestley 1981) is used to smooth the noise-corrupted spectra, by which the spectral amplitudes within a specified frequency bandwidth are smoothed. Again, SSRM is used in the frequency domain identification procedure as summarized in Figure 7.

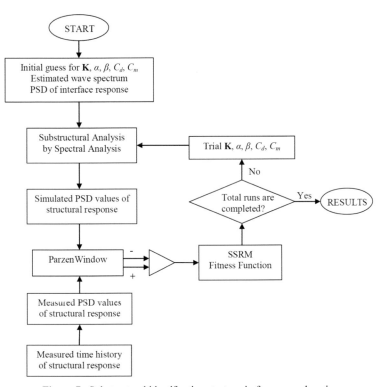

Figure 7. Substructural identification strategy in frequency domain.

4.2 Parameter Identification of Substructures

Similar to time domain identification, sensitivity study in frequency domain is firstly conducted (Wang 2012). It is also shown that rotational response is more sensitive to leg stiffness change in S1 than linear responses. Therefore, the rotational accelerations of internal nodes in S1 are selected to define the fitness function for frequency domain identification. The sensitivity study for S2 demonstrates that rotational acceleration is more sensitive to leg stiffness changes. Horizontal and vertical accelerations are more sensitive to spudcan stiffness changes than rotational accelerations. Moreover, it is noted that dynamic response is more sensitive for spudcan fixity especially for vertical stiffness and rotational stiffness. It indicates that PSD values of horizontal, vertical, and rotational accelerations are all sensitive to structural stiffness changes in S2. Fitness function should involve different types of structural responses for S2 identification. After several tests of response combinations, one set of structural response is determined: horizontal responses of internal nodes excluding spudcan node are used in the fitness function and additional two pairs of vertical and rotational responses at 3rd and 5th nodes are included.

Substructural identification is carried out with SSRM parameters, search limits and noise levels same as those in time domain identification (see Table 2). The identification results of ten tests are obtained and presented in terms of ratios between identified values of mean parameters and exact values, as shown in Figure 8. The identification errors of mean parameters are also presented in Table 5. The errors of mean parameters based on frequency domain identification are less than 15% for 10% noise level and less than 10% for 5% noise level respectively. In general, time domain identification performs better than frequency domain identification. A possible explanation is that only the spectral amplitudes are used in the frequency domain approach and the phase angle information is excluded. Besides, conversion from time histories to spectral estimates may add some computational errors.

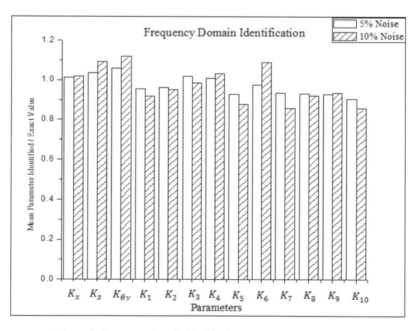

Figure 8. Frequency domain identification results in numerical study.

Table 5. Errors of mean parameters (%) in frequency domain.

Noise Level	K_x	K_z	$K_{\theta x}$	K_1	K_2	K_3	K_4	K_5	K_6	K_7	K_8	K_9	K_{10}
5%	1.30	3.57	5.99	4.35	3.89	1.88	1.00	7.02	2.38	6.58	6.80	7.28	9.61
10%	1.77	9.39	11.82	7.97	4.77	1.48	3.08	12.13	8.86	14.22	7.77	6.44	14.36

4.3 Damage Detection in Frequency Domain

Damage detection can be carried out in frequency domain based on accurate parameter identification. The damage definition in frequency domain identification is the same as that in time domain identification, as expressed in Equation (23). The damage detection strategy is to compute the stiffness changes between the damaged structure and the undamaged structure, based on the available measurements before and after damage. The structure mass is assumed to be known and unchanged. The identified parameters of the undamaged structure are included in the initial population for the identification of the damaged structure, same as damage detection in time domain. The search limits for unknown parameters are also set as half to double of the actual parameters of the undamaged structure. Considering the critical parts, for S1 element K_8 near the water surface is considered to be the damage location and for S2 the damage is considered in rotational stiffness of the spudcan fixity.

In the numerical study, I/O noise is set at 5%. The identified parameters of the undamaged structure in frequency domain at 5% noise level are used as good starting points for identification of the damaged structure. Two damage cases are simulated and studied. One case is 20% stiffness loss of K_8 in S1 and the other case is 20% stiffness loss of in S2. In total, 10 tests of damage detection are carried out in frequency domain for each case. The numerical results of damage detection in frequency domain are derived and listed in Table 6.

In Case 1, the stiffness loss of K_8 in S1 is clearly identified as highlighted in bold in Table 6 (a). The largest difference between the detected damage and the exact damage is less than 5% and the mean damage of K_8 detected is –21.65% which is in good agreement with the exact damage of 20%. Similarly, as shown in Table 6(b) for Case 2, the damage in rotational stiffness of the spudcan fixity in S2 is also well-detected as the results in bold. The largest difference between the detected damage and the exact damage is less than 3% and the mean damage of K_r detected is –19.92%, very close the exact damage of 20%. Besides, some false damages are also identified for other stiffness values, but much smaller than the damages detected in K_8 for Case 1 and in for Case 2. Compared with damage detection in time domain, the false damage values increase but the mean values are still very small as less than 3% which is less than the 5% I/O noise level. Similar with damage

Table 6. Damage detection results in frequency domain.

(a) Case 1: identified damage (%) in S1

	1	2	3	4	5	6	7	8	9	10	Mean
K_6	−5.20	−6.79	−1.11	3.84	−4.02	−2.99	6.13	4.87	−4.70	2.93	−0.70
K_7	−6.84	−5.82	−4.35	−0.94	−6.23	5.25	0.01	−3.78	−0.64	−3.37	−2.67
K_8	**−24.36**	**−22.86**	**−22.44**	**−19.74**	**−19.56**	**−20.97**	**−19.33**	**−22.51**	**−21.69**	**−23.08**	**−21.65**
K_9	−2.83	0.02	−1.84	−0.80	−5.13	7.14	0.03	0.99	−3.26	−5.89	−1.16
K_{10}	−0.48	1.14	−0.25	−3.01	4.05	4.73	6.01	6.72	1.59	0.79	2.13

(b) Case 2: identified damage (%) in S2

	1	2	3	4	5	6	7	8	9	10	Mean
K_1	0.69	0.02	−0.12	2.52	1.80	5.76	2.86	0.09	−2.46	4.27	1.54
K_2	−2.26	0.44	−2.89	−2.66	−0.81	1.54	0.01	−4.17	−2.52	0.82	−1.25
K_3	−1.01	0.15	0.95	−0.72	−1.69	0.33	−4.01	−0.28	5.93	−5.24	−0.56
K_4	0.45	0.02	1.35	−4.23	−2.13	−5.03	0.03	−2.82	0.44	−2.94	−1.49
K_5	−0.57	1.21	1.68	1.77	3.20	0.55	0.02	1.56	−3.28	2.38	0.85
K_x	−1.96	−1.70	−1.43	−0.25	−2.45	−1.80	−1.08	−1.78	6.74	1.19	−0.45
K_z	−0.08	0.02	0.22	−0.71	0.36	3.30	0.03	−2.66	6.70	−2.19	0.50
$K_{\theta y}$	**−16.92**	**−21.92**	**−20.49**	**−20.20**	**−21.18**	**−19.96**	**−18.22**	**−17.85**	**−18.12**	**−18.03**	**−19.29**

detection in time domain, it is noted that for different tests in frequency domain the damage values vary as positive or negative except for the real damage location, at which the damage values are always negative clearly pointing to the occurrence of the damage. The numerical results demonstrate that accurate damage detection in frequency domain can be derived in terms of quantity and location. Therefore, the proposed identification strategy in frequency domain can be extended to damage detection in a non-destructive way.

5. Experimental Study for Support Fixity Identification of Jack-up System

Numerical study for system identification of jack-up model has been carried out in both time domain and frequency domain. The identification results are acceptable even under noisy conditions. To further test and verify the effectiveness of the proposed identification strategies, a laboratory experimental study is also carried out. Since spudcan fixity is critical for jack-up platform, the experimental study will focus on the identification of support fixity.

5.1 Laboratory Model

A small-scale laboratory model, similar to the jack-up used in the numerical study, was designed and fabricated. The overall length dimensions are scaled by a scale factor $C = 1:210$. Since the flexural rigidity is more important, the moment of area of the leg is scaled by C^4. The cross-sectional area of the leg is too small to be scaled by C^2. Thus, only the moment of area of the leg is scaled. To simplify the experimental model, rotational stiffness of the support is mainly considered and designed by using a particular spring support due to its significance. The rotational stiffness of the support is modeled to be comparable to the leg flexural rigidity based on practical conditions. The purpose of the experimental study is not to achieve full similitude (which is difficult due to the small scale of the experimental model) but to demonstrate the proposed identification strategies. The main structural parameters of the designed model are listed in Table 7. The main components of the experimental model are summarized in Table 8, according to the dimensional drawing as shown in Figure 9. The experimental model includes the rotational

Table 7. Structural parameters for experimental model.

Structural Parameters	Prototype Model	1:210 Model	Built Model
Length of leg L_{leg} (m)	120	0.571	0.571
Dist. from FW leg to AF legs H(m)	42	0.2	0.2
Young's modulus of leg E(Mpa)	200000	69000	68670
Moment of area of single leg I(m^4)	10.462	5.379×10^{-9}	5.271×10^{-9}
Cross-sectional area of single leg A(m^2)	0.444	1.007×10^{-5}	9.766×10^{-5}
Leg flexural factor EI/L_{leg}(Nm)	1.744×10^{10}	6.536×10^2	6.339×10^2
Support rotational stiffness K_θ(Nm/rad)	1.780×10^{10}	6.671×10^2	6.520×10^2

Table 8. Main Components of Experimental Model (according to Figure 9).

No.	Item	Material	Dimension	Quantity
	Hull:			
1	Triangular block	Aluminum	Details see Figure 9	1
	Leg:			
2	Tube	Aluminum	$D_o = 22.225$ mm, $t = 1.5$ mm, $L = 571$ mm	3
	Spring Support:			
3	Top plate	Steel	$t = 5$ mm, details see Figure 9	3
4	Spring	Steel	$L = 75$ mm, $D_o = 30$ mm, $t = 5$ mm, Pitch = 10 mm, 8.5 coils	12
5	Bearing	Steel	$D = 30$ mm, $d = 12$ mm, Width = 10 mm	6
6	Bearing holder	Steel	$t = 5$ mm, details see Figure 9	3
7	Pivot	Steel	$d = 12$ mm, $L = 40$ mm	3
8	Bracket	Steel	$t = 5$ mm, details see Figure 9	6

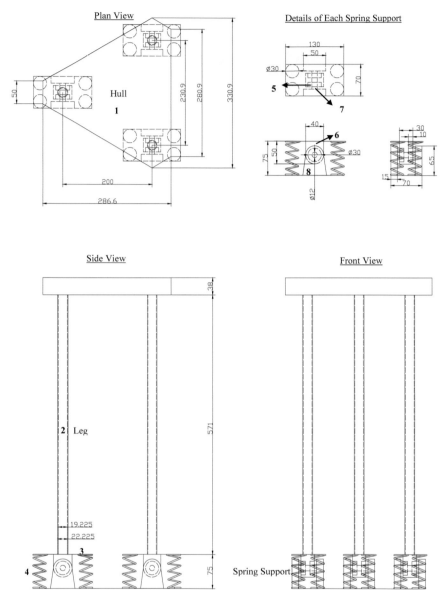

Figure 9. Layout of experimental jack-up model.

stiffness of the support by means of springs and bearings and the details are also drawn in Figure 9.

5.2 Experimental Tests

In order to verify the dynamic properties of the model, preliminary tests are carried out before the set-up of the complete jack-up model (Wang 2012). Rotational stiffness for each spring support and the combined leg stiffness are measured in static tests. The stiffness values can be used as a benchmark for validating the results derived from the proposed substructural identification strategies (i.e., to compare the identified values with the benchmark values). As shown in Figure 10, the jack-up model is analyzed by finite element method and the lower part of a single leg is considered as the substructure to be identified. Impact tests are also carried out and show that the measured fundamental frequency is almost the same as the computed one based on measured stiffness.

For structural identification, dynamic tests are carried out by applying Gaussian random force series with a bandwidth of (0–1000 Hz). The force signal is generated by a signal generator and then passed through a power amplifier in order to produce sufficient power for the electromagnetic shaker (Model Labworks ET-126B). The force generated by the shaker is applied to the structure via a connecting rod to one side of the hull (see Figure 11), and then measured by a force sensor (PCB-208C02). A tri-axial linear accelerometer (Dytran-3293A) and a Gyro sensor (SD-740) are used to measure the interface response, i.e., horizontal, vertical accelerations and rotational velocity (rotational acceleration is derived by further data processing). Substructural response at internal nodes is measured

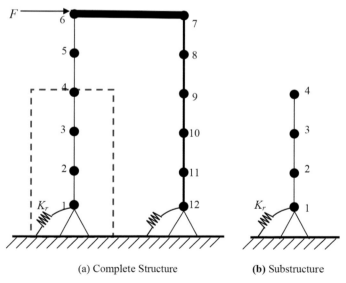

(a) Complete Structure   **(b)** Substructure

Figure 10. Numerical model for experimental study.

by two single-axial linear accelerometers (Dytran-3055B3). The instrumented jack-up model is shown in Figure 12. Ten different series of random force signals (F1~F10) are applied in dynamic tests. The response signals are recorded by a digital oscilloscope at a sampling rate of 2 thousand samples per second. The noise level of the measurements is around 15%, which further serves as a test of the proposed strategies with regard to noise effects.

Figure 11. Shaker connection detail.

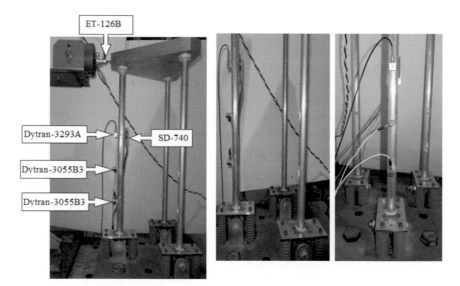

Figure 12. Dynamic tests for experimental jack-up model.

5.3 Substructural Identification of Experimental Model

In the experimental study, the support fixity is set as the parameter to identify and damping coefficients are also treated as unknowns. The mass and leg stiffness determined by static tests are used. As in the numerical study, the initial guess is randomly generated within a range from half to double of benchmark values. Based on the measurements of the interface, horizontal accelerations of internal nodes can be estimated in substructural analysis and then compared with measurements in the fitness function.

System identification is carried out in both time domain and frequency domain. For time domain identification, all the 4096 data points of structural response with a time step of 0.5 ms are used for each force signal and the filter window is set between 200 Hz and 500 Hz. For frequency domain identification, 2048 data points of structural response are selected for each force signal and then 14,336 tailing zeros are added. In total, 16,384 data points of structural response are prepared, which are then transformed to frequency domain by FFT. Thus, the frequency resolution is about 0.122 Hz. The frequency range used for system identification is set between 200 Hz and 500 Hz.

The identification results of support fixity for ten different force signals are derived and then compared with the benchmark value. In order to clearly demonstrate the identification results, the ratios between identified values and benchmark value are computed and shown in Figure 13. The identification errors

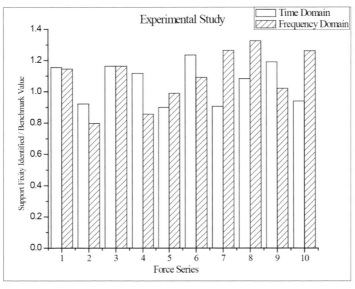

Figure 13. Identification results in experimental study.

125

between identified values and static measurements together with the mean errors are also summarized in Table 9. Again, it is advantageous to average out the values identified in different tests; the errors of the mean rotational stiffness of the support is about 6% for time domain identification and 9% for frequency domain identification.

As shown in Figure 13, the identified rotational stiffness of the support is close to the static measurement. It indicates that time domain identification and frequency domain identification are both applicable and effective for the experimental study. The identification results are fairly accurate especially for time domain identification, and the largest percentage difference is about 23% as shown in Table 9. Frequency domain identification results have more variation because frequency domain identification involves comparison between measured PSD values derived by FFT and simulated PSD values derived from spectral analysis, which might bring some numerical errors from data transformation. However, the errors for frequency domain identification are not significant with largest percentage difference of about 32% as shown in Table 9. Moreover, in Table 9, the identification errors are acceptable for both time domain identification and frequency domain identification. The accuracy of support fixity identification is adequate for practical requirement.

It is shown that the identification strategies proposed based on GA method is practically effective and accurate for jack-up model. Moreover, the error of mean parameter is much smaller than the mean error based on different force signals. The mean parameter can predict the support fixity within less than 10% error. It is recommended to average the identification parameters in practice.

Table 9. Absolute errors (%) of support stiffness (K_r) identification in experimental study.

	F1	F2	F3	F4	F5	F6	F7	F8	F9	F10	Error of Mean K_r
TD	15.53	7.82	16.26	11.61	10.10	23.43	9.47	8.23	18.93	6.18	6.04
FD	14.56	20.36	16.31	14.43	1.04	9.00	26.29	32.44	2.00	25.97	9.07

Note: TD and FD represent time domain and frequency domain respectively.

6. Conclusions and Recommendations

System identification algorithms have been applied to onshore structures for parameter identification, damage detection and structural health monitoring for years. Nevertheless, relatively few studies have been reported on offshore structures which present more challenges than onshore structures. To this end, the primary objective of this research is to develop robust and effective identification strategies for offshore structural systems, focusing on jack-up platform as an example of application. Leg stiffness and spudcan fixity are the key parameters to identify. Two effective identification strategies are proposed in time domain and in frequency domain based on genetic algorithms (GA) and substructural identification. In addition, an experimental study is conducted to substantiate the numerical study and enhance the value of this research.

6.1 Conclusions

The procedures of dynamic analysis for complete structure and substructure are introduced in both time domain and frequency domain. Taking a jack-up platform in the North Sea as an illustration example, a single leg modeled by frame elements is studied as two substructures considering the critical parts. The inclusion of spudcan fixity is beneficial for design and operation in deeper water. One challenge for time domain analysis is that the initial condition is difficult to determine or measure. Nevertheless, the transient response due to initial conditions decays rapidly with time. This is an important point which would overcome the problem of unknown initial conditions in time domain identification. Alternatively, spectral analysis is more appropriate to avoid considering initial conditions. The numerical results show that structural response derived from time domain analysis is in good agreement with that derived from frequency domain analysis in terms of power spectral density. Most importantly, substructural analysis is shown to be consistent with complete structural analysis and provides the basis for substructural identification (Wang 2012).

Two GA-based substructural identification strategies with multiple novel ideas are proposed in time domain and in frequency domain to address the challenges faced in the identification of offshore jack-up systems. Substructural identification can improve efficiency and accuracy significantly by reducing the system size and the number of unknown parameters. It permits the study of critical parts including leg and spudcan. Structural damping and hydrodynamic coefficients are considered as unknown in the identification as well. Besides, 2D frame jack-up model is sufficient as a novel way of substructuring because the plane of consideration for the selected leg can be set arbitrarily and the measurements are set in the corresponding plane. In the GA-based search, the search limits for unknown parameters are reasonably wide with an initial guess of unknown parameters randomly generated. Good initial guess is therefore not needed, unlike some other identification methods. GA has the flexibility of using any response in the fitness function and has better global search capacity due to probabilistic rules and population-to-population searches. In this research, improved fitness functions are developed, with regard to normalized structural response, which enables the identification of complex dynamic system based on different types of measurements (linear and rotational) in the fitness computation.

127

To address the challenges of unknown wave loading and unknown initial conditions requires different novel ideas in time domain identification and in frequency domain identification. In time domain identification, an output-only algorithm is adopted to avoid wave force measurements. Since transient oscillations decay rapidly by damping effects, the transient state due to unknown initial conditions is excluded in the fitness computation for time domain identification. In frequency domain identification, no initial conditions are involved, which is an inherent advantage of spectral analysis. To deal with unknown wave loading, the frequency domain method takes advantage of the fact that operations on the hull excite a dynamic response in higher frequency range than wave loading. If not sufficient, it is proposed to apply an additional random force on the hull to generate a greater substructural response (according to the natural frequencies of the substructures). The fitness function is selectively defined to cover the main frequencies of the substructure but away from the dominant (low) frequency range of wave loading. As such, unknown wave forces are not involved in the proposed frequency domain method.

In the numerical study of stiffness parameter identification, noise effects are considered at three different levels (0%, 5%, and 10%). Ten different tests of system identification for each substructure are carried out for each noise level in time domain and in frequency domain respectively. Numerical results are obtained for identification of stiffness parameters including leg stiffness and spudcan fixity. As the noise level increases, the identification errors are expected to increase, but the identification results under noisy conditions are also acceptably accurate. The error range of spudcan fixity identification is similar to that of leg stiffness identification, within about $\pm$ 10% for 5% noise and about $\pm$ 20% for 10% noise. Thus, the inclusion of both leg stiffness and spudcan fixity in system identification is achievable with good accuracy. It is recommended to average the identified parameters from different tests to achieve better identification results. In general, time domain identification performs better than frequency domain identification, with errors of the mean parameters less than 8% for 10% noise level because the conversions of the measurements from time domain to frequency domain may introduce numerical errors.

Damage detection is also carried out based on identification of stiffness changes before and after damage. The identified parameters of the undamaged structure are included in the initial guess for the identification of the damaged structure. I/O noise level is set at 5% and two damage cases with 20% loss of single stiffness in leg and in spudcan are studied. Ten different tests of damage detection are carried out for each damage case in time domain and infrequency domain respectively. The results show that damages can be consistently and accurately detected based on identified damage values obtained from different tests. The mean identified damages are very close to the exact damages with less than 2% difference.

128

To reinforce the findings, a laboratory experimental study of a small-scale jack-up model is also carried out. The support fixity in terms of rotational stiffness is the key parameter of identification. Based on the measurements of dynamic response, substructural identification is carried out by the proposed identification strategies. It is shown that the identified rotational stiffness of the support is close to the benchmark value obtained by static test measurement. The errors of the mean parameters for both time domain identification and frequency domain identification are less than 10%. The experimental results further demonstrate that the proposed identification strategies are accurate and effective.

6.2 Recommendations

It is highly anticipated that the systematic methodologies developed in this research can be extended to other applications of offshore systems such as floating platforms, pipelines, risers, mooring and subsea systems. Based on both numerical and experimental studies, it is shown that there are some potential areas for future work.

For jack-up platform, different substructures may be considered and studied to identify other structural parameters such as leg-hull connections which are difficult to model. Influenced by jacking systems, leg-hull connections can provide bending moments after installation, which is useful for platform stability. The identification of leg-hull connections will result in accurate model calibration. Besides, an improved numerical model including nonlinear aspects in both structural response and spudcan fixity may be developed for more detailed dynamic analysis. Also, it is possible to extend to 3D model when analyzing some global response such as torsion of the entire platform or the legs. However, when complex numerical model is used, computer time will be increased accordingly. Thus, the identification efficiency will become a critical issue. Incorporating GA-based search and some local search techniques may be considered to improve computing efficiency. Besides, parallel computing of system identification is expected to have a further reduction of computer time. Moreover, it requires further specific research for the extension of the methodology and system developed in this research to other offshore structural systems.

For an experimental study on support identification, horizontal and vertical stiffness of the support may be included in further research. The more complex model is developed, the more DOFs are involved. Correspondingly, improvements in sensors especially for high-performance angular accelerometers are required. The proposed identification strategies may be applied to operating platforms at sites. Data quality of practical measurements is important for the accuracy of system identification, regardless of time domain identification or frequency domain identification. Therefore, many practical issues need to be resolved in particular engineering situations.

Acknowledgement

This research is supported by the Research Start-up Foundation for Advanced Talent of Jiangsu University, Six Talent Peaks Project in Jiangsu Province (No.: 2017-KTHY-010), and the research grants (No.: R-264-000-226-305 and R-264-000-226-490) funded by the Agency for Science, Technology and Research as well as Maritime and Port Authority of Singapore.

References

Alrashidi, M.R. and EI-Hawary, M.E. (2006). A survey of particle swarm optimization applications in power system operations. Electric Power Components and Systems, 34: 1349–1357.
Barltrop, N.D.P. and Adams, A.J. (1991). Dynamics of Fixed Marine Structures, Third Edition. Butterworth-Heinemann Ltd., UK.
Bayissa, W.L. and Haritos, N. (2009). Structural damage identification using a global optimization technique. International Journal of Structural Stability and Dynamics, 9: 745–763.
Bienen, B. and Cassidy, M.J. (2006). Advances in the three-dimensional fluid-structure-soil interaction analysis of offshore jack-up structures. Marine Structures, 19: 110–140.
Bretschneider, C.L. (1959). Wave variability and wave spectrum for wind generated gravity waves. US Army Corp. of Engineers. Beach Erosion Board, Tech Memo: No. 18.
Carden, E.P. and Fanning, P. (2004). Vibration based condition monitoring: a review. Structural Health Monitoring, 3: 355–377.
Cassidy, M.J., Houlsby, G.T., Hoyle, M. and Marcom, M.R. (2002). Determining appropriate stiffness levels for spudcan foundations using jack-up case records. pp. 307–318. *In*: The Proceedings of the 21st International Conference on Offshore Mechanics and Arctic Engineering (OMAE), Oslo, Norway.
Cassidy, M.J., Vlahos, G. and Hodder, M. (2010). Assessing appropriate stiffness levels for spudcan foundations on dense sand. Marine Structures, 23: 187–208.
Chakrabarti, S.K. (1987). Hydrodynamics of Offshore Structures, WIT Press, Southampton.
Chakrabarti, S.K. (2005). Handbook of Offshore Engineering, Elsevier Science, Oxford, UK.
Chang, P.C., Flatau, A. and Liu, S.C. (2003). Review paper: health monitoring of civil infrastructure. Structural Health Monitoring, 2: 257–267.
Chen, J. and Li, J. (2004). Simultaneous identification of structural parameters and input time history from output-only measurements. Computational Mechanics, 33: 365–374.
Clough, R.W. and Penzien, J. (1993). Dynamics of Structures. 2nd ed. McGraw-Hill. New York.
Dean, R.G. and Dalrymple, R.A. (1984). Water Wave Mechanics for Engineers and Scientists, Prentice-Hall Inc., New Jersey, USA.
Denton, A.A. (1986). Introduction. pp. 1–7. *In*: Boswell, L.F. (ed.). The Proceeding of 1st International Conference on the Jack-up Drilling Platform, Design and Operation, London, UK.
Dessi, D. and Carmerlengo, G. (2015). Damage identification techniques via modal curvature analysis: overview and comparison. Mechanical Systems and Signal Processing, 52-53: 181–205.
Ewins, D.J. (2000). Modal Testing: Theory, Practice and Application. Research Studies Press, London.
Fan, W. and Qiao, P.Z. (2011). Vibration-based damage identification methods: a review and comparative study. Structural Health Monitoring, 10: 83–111.
Feng, M.Q. and Bahng, E.Y. (1999). Damage assessment of jacketed RC columns using vibration test. Journal of Structural Engineering, 125: 265–271.
Friswell, M.I. (2007). Damage identification using inverse methods. Philosophical Transactions of the Royal Society A on Structural Health Monitoring, 365: 393–410.

Ghanem, R. and Shinozuka, M. (1995). Structural system identification I: theory. Journal of Engineering Mechanics, 121: 255–263.

Hao, H. and Xia, Y. (2002). Vibration-based damage detection of structures by genetic algorithm. Journal of Computing in Civil Engineering, 16: 222–229.

Hasselmann, K., Barnett, T.P., Bouws, E. et al. (1973). Measurements of wind wave growth and swell decay during the JONSWAP project. Deut Hydro, Z12: 1–95.

Hsieh, K.H., Halling, M.W. and Barr, P.J. (2006). Overview of vibrational structural health monitoring with representative case studies. Journal of Bridge Engineering, 11: 707–715.

Humar, J., Bagchi, A. and Xu, H.P. (2006). Performance of vibration-based techniques for the identification of structural damage. Structural Health Monitoring, 5: 215–241.

Karthigeyan, V. (2009). Dynamic shear modulus of soils foundation stiffness and damping for seismic analysis of jack-ups. *In*: Proceedings of the 12th International Conference on the Jack-up Platform, London, UK.

Koh, C.G., Hong, B. and Liaw, C.Y. (2000). Parameter identification of large structural systems in time domain. Journal of Structural Engineering, 126: 957–963.

Koh, C.G., Hong, B. and Liaw, C.Y. (2003). Substructural and progressive structural identification methods. Engineering Structures, 25: 1551–1563.

Koh, C.G. and Perry, M.J. (2010). Structural Identification and Damage Detection using Genetic Algorithms. Taylor & Francis, London.

Koh, C.G., See, L.M. and Balendra, T. (1991). Estimation of structural parameters in the time domain: a substructure approach. Earthquake Engineering and Structural Dynamics, 20: 787–801.

Lu, Z.R. and Law, S.S. (2006). Force identification based on sensitivity in time domain. Journal of Engineering Mechanics ASCE, 132: 1050–1056.

Luh, G.C. and Wu, C.Y. (1999). Non-linear system identification using genetic algorithms. Journal of Systems and Control Engineering, 213: 105–118.

Maia, N.M.M. and Silva, J.M.M. (2001). Theoretical and Experimental Modal Analysis. Research Studies Press, Hertfordshire, UK.

Mangal, L., Idichandy, V.G. and Ganapathy, C. (2001). Structural monitoring of offshore platforms using impulse and relaxation responses. Ocean Engineering, 28: 689–705.

Mommaas, C.J. and Dedden, W.W. (1989). The Development of a stochastic non-linear and dynamic jack-up design and analysis method. Marine Structures, 2: 335–363.

Morison, J.R., O'Brien, M.P., Johnson, J.W. and Schaaf, S.A. (1950). The force exerted by surface waves on piles. Petroleum Transactions, 189: 149–154.

Najafian, G. (2007a). Application of system identification techniques in efficient modeling of offshore structural response. Part I: Model development. Applied Ocean Research, 29: 1–16.

Najafian, G. (2007b). Application of system identification techniques in efficient modeling of offshore structural response. Part II: Model validation. Applied Ocean Research, 29: 17–36.

Nataraja, R., Hoyle, M.J.R., Nelson, K. and Smith, N.P. (2004). Calibration of seabed fixity and system damping from GSF Magellan full-scale measurements. Marine Structures, 17: 245–260.

Nelson, K., Stonor, R.W.P. and Versavel, T. (2001). Measurements of seabed fixity and dynamic behaviour of the Santa Fe Magellan jack-up. Marine Structures, 14: 451–483.

Parzen, E. (1962). Stochastic Processes. Holden-Day, San Francisco, USA.

Perra, R. and Torres, R. (2006). Structural damage detection via modal data with genetic algorithms. Journal of Structural Engineering, 132: 1491–1501.

Perry, M.J. and Koh, C.G. (2008). Output-only structural identification in time domain: numerical and experimental studies. Earthquake Engineering and Structural Dynamics, 37: 517–533.

Pierson, W.J. and Moskowitz, L.A. (1964). A proposed spectral form for fully developed wind seas based on the similarity theory of S.A. Kitaigorodskii. Journal of Geophysical Research, 69: 5181–5190.

Poulos, H.G. and Davis, E.H. (1974). Elastic Solutions for Soil and Rock Mechanics. John Wiley, New York.

Priestley, M.B. (1981). Spectral Analysis and Time Series. Academic Press, London.

Senner, D.W.F. (1993). Analysis of long term jack-up rig foundation performance. pp. 691–716. *In*: Proceeding of Conference on Advances in Underwater Technology, Ocean Science and Offshore Engineering, Offshore Site Investigation and Foundation Behaviour, London, United Kingdom.

Shinozuka, M., Yun, C.B. and Vaicaitis, R. (1977). Dynamic analysis of fixed offshore structures subjected wind generated waves. Journal of Structural Mechanics, 5: 135–146.

Shyam Sunder, S. and Sanni, R.A. (1984). Foundation stiffness identification for offshore platforms. Applied Ocean Research, 6: 148–156.

SNAME T&R 5-5A (2002). Guidelines for Site Specific Assessment of Mobile Jack-Up Units. Society of Naval Architects and Marine Engineers, New Jersey, USA.

Springett, C.N., Stonor, R.W.P. and Wu, X. (1996). Results of a jack-up measurement programme in the North Sea and their comparison with the structural analysis. Marine Structures, 9: 53–70.

Temperton, I., Stonor, R.W.P. and Springett, C.N. (1999). Measured spudcan fixity: analysis of instrumentation data from three North Sea jack-up units and correlation to site assessment procedures. Marine Structures, 12: 277–309.

Trinh, T.N. and Koh, C.G. (2012). An improved substructural identification strategy for large structural systems. Structural Control and Health Monitoring, 19: 686–700.

Vlahos, G., Cassidy, M.J. and Martin, C.M. (2008). Experimental investigation of the system behavior of a model three legged jack-up on clay. Applied Ocean Research, 30: 323–337.

Vlahos, G., Cassidy, M.J. and Martin, C.M. (2011). Numerical simulation of pushover tests on a model jack-up platform on clay. Geotechnique, 61: 947–960.

Wang, X.M. (2012). System Identification of Jack-up Platform by Genetic Algorithms. PhD Thesis, National University of Singapore.

Wang, X.M., Koh, C.G. and Zhang, J. (2014). Substructural identification of jack-up platform in time and frequency domains. Applied Ocean Research, 44: 53–62.

Yang, J.N., Pan, S. and Lin, S. (2004). Identification and tracking of structural parameters with unknown excitations, 4189–4194. In the Proceedings of the 2004 American Control Conference, Boston, USA.

Yun, C.B. and Shinozuka, M. (1980). Identification of nonlinear structural dynamic systems. Journal of Structural Mechanics, 8: 187–203.

Zhang, J., Koh, C.G., Trinh, T.N., Wang, X.M. and Zhang, Z. (2012). Identification of jack-up spudcan fixity by an output-only substructural strategy. Marine Structures, 29: 71–88.

Zhang, Z., Koh, C.G. and Duan, W.H. (2010). Uniformly sampled genetic algorithm with gradient search for structural identification - Part II: Local search. Computers & Structures, 88: 1149–1161.

Zubaydi, A., Haddara, M.R. and Swanmidas, A.S.J. (2002). Damage identification in a ship's structure using neural networks. Ocean Engineering, 29: 1187–1200.

6

Wave Energy Converter Arrays for Electricity Generation with Time Domain Analysis

Fuat Kara

1. Introduction

The current development pace of wave energy converters indicates the possibility of the deployment of these converters as arrays at commercial scale. The accurate predictions of wave loads, motion characteristics, and power requirements are critically important for the design of these devices which are in sufficiently close proximity to experience significant hydrodynamic interactions. The oscillation of each body radiates waves assuming that other bodies are not present. Some of these radiated waves that can be considered as incident waves interact with the bodies of the array causing diffraction phenomena while others radiate to infinity. The fluid response between arrays can affect overall power generation and could increase or decrease power generation compared to an isolated device. The power generation due to hydrodynamic interaction depends on separation distance, geometrical layout, direction of the incident wave, geometry in the array, incident wave length, mooring configurations, control strategies, etc.

Cranfield University, MK43 0AL, UK.
E-mail: f.kara@cranfield.ac.uk

The pioneer work of (Budalk 1977) on wave energy converter arrays introduced the point absorber approximation in which the response amplitude are considered as equal for all devices and optimal power absorption are independent from device geometry. Moreover, the characteristic dimensions (e.g., diameter) of the devices are considered small in terms of incident wave length. This approximation implicitly means that wave diffraction is not significant and can be ignored (Thomas and Evans 1981). In these studies, the overall absorbed power value which increases and decreases with the wave impact is measured by interaction factor q-factor (q > 1 is for power increase and q < 1 is for power decrease). The interaction factor q-factor is measured as the ratio of power from an array to n times power from an isolated device. This q-factor is used to optimize the array layout in order to get maximum power (Fitzgerald and Thomas 2007). One of the important finding from this work (Fitzgerald and Thomas 2007) was that the average value of q-factor is unity when overall heading is taken into account. This implicitly means that the power absorption is constructive in some headings while it is destructive in other headings (Wolgamot et al. 2012).

The restriction of point absorber approximation related to diffraction waves was removed by the use of plane wave analysis in which interactions of diverging waves considered as plane waves between devices are taken into account while the near-field waves (or evanescent waves) effects are ignored implying separation distance between devices is large relative to wavelength (Mclever and Evans 1984, Simon 1982, Singh and Babarit 2014, Spring and Monkmeyer 1974). The restriction on separation distance between devices or exclusion of near-field waves was overcome by the use of multiple scattering methods in which the superposition of incident wave potential, diverging and near-field waves, and radiated waves by the oscillation of devices are taken into account. In this way, the wave field around devices can be represented accurately (Mavrakos 2004, Ohkusu 1972, Ohkusu 1974, Tweresky 1952). As an accurate solution requires high number of diffracted and radiated wave superposition with iteration; this process increases the computational time significantly (Linton and Mclever 2001).

The restriction on the computational time was avoided by the use of the direct matrix method in which the multiple scattering prediction are combined with a direct approximation (Kagemoto and Yue 1986) and unknown wave amplitudes are predicted simultaneously rather than iteratively. As the numerical results of this approach, which exactly depended on infinite summation truncation, were very accurate compared to other numerical approximations, this method was applied to many different engineering problems including near trapping problem in large arrays (Maniar and Newman 1997), very large floating structures (Kagemoto and Yue 1993, Kashiwagi 2000), tension-leg-platforms (Yilmaz 1998), and wave energy converters (Child and Venugopal 2007, Child and Venugopal 2010).

In addition to the above exact formulations, the numerical tools to predict hydrodynamic interactions for multi-bodies are studied extensively by many researchers including (van't and Siregar 1995) who used the strip theory in which the hydrodynamic interactions are considered as two-dimensional flow. The unified theory was used to overcome the low frequency limitations of strip theory (Breit and Sclavounos 1986, Kashiwagi 1993, Ronaessm 2002). These two-dimensional approaches provide poor predictions as the hydrodynamic interactions including separation distances between the bodies are neglected in the calculations.

As the hydrodynamic interactions are inherently three-dimensional, numerical approximations need to be used for accurate prediction of the wave loads and motions over array systems as three-dimensional effects play a significant role in the dissipation of wave energy between bodies in arrays. The hydrodynamic interaction effects are automatically taken into account as each discretized panel would have its influence on all other panels in three-dimensional numerical models. The viscous Computational Fluid Dynamics (CFD) methods for full fluid domain or viscous CFD in the near field and inviscid CFD in the far field can be used for the prediction of three-dimensional non-linear flow field due to incident waves. However, the required computational time to solve these kinds of problems is not suited for practical purposes yet (Wei et al. 2013, Westphalen et al. 2014, Yu and Hand 2013). In these works, single wave energy converter is used with the order of 10^6 cells and the solution for single WEC requires 24–72 hrs to compute approximately 10 s.

An alternative approach to a viscous solution is the three-dimensional potential flow approximation to solve the hydrodynamic interactions. The computational time of potential (or inviscid CFD) which neglects the viscous effect is much less than viscous CFD and is used to predict the hydrodynamic loads over floating single bodies and arrays. The prediction of the effects of three-dimensional hydrodynamic interactions on arrays can be obtained using three-dimensional frequency and time domain approaches. Two kinds of formulations were used for this purpose. These are Green's function formulation (Kara 2000, King 1987, Liapis 1986) and Rankine type source distribution (Bertram 1990, Kring and Sclavounos 1991, Nakos and Scravonous 1990, Nakos et al. 1993, Xiang and Faltinsen 2011).

The Green function's approach satisfies the free surface boundary condition and condition at infinity automatically, and only the body surface needs to be discretized with panels, while the source and dipole singularities are distributed discretizing both the body surface and a portion of the free surface in Rankine type formulation. The main disadvantage of Rankine type source distribution is the stability problem for the numerical implementation since the radiation condition or condition at infinity is not satisfied exactly. The requirement of the discretization of some portion of the free surface using quadrilateral or triangular elements increases the computational time substantially. The time domain and the frequency domain results are related by the Fourier Transform in the context of the linear theory.

There are many different radiation/diffraction numerical codes that are used for the prediction of wave energy converter variables such as WAMIT (WAMIT User Manual 2012), Aquaplus (Delhommeau 1993), AQWA (ANSYSA 2013). WAMIT (WAMIT User Manual 2012) is the most widely used commercial programme and uses frequency domain Green function to predict hydrodynamics loads over single bodies and arrays and is used by many researchers for many different purposes including optimizing the damping characteristics of wave energy converters in the arrays, increasing the capture width of a line absorber with cylindrical floats, and analysing the motions of a floating platform with several absorber attached (Bellew et al. 2009, De Backer et al. 2009, Stansby et al. 2015, Taghipour and Moan 2008, Child and Cruz 2011). Other non-commercial numerical approaches are used by many other researchers to predict the effects of hydrodynamic interactions on absorbed power, captured width, separation distances, directionality, interaction factors, efficiency of the numerical methods in the case of large number of wave energy converters in the array, and optimum control (Babarit 2010, Borgarino et al. 2012, Justino and Clement 2003, Kara 2010, Babarit et al. 2010, Babarit 2013, Folley et al. 2012, Konispoliatis and Mavrakos 2013, Mclever 2002, Nader et al. 2014, Siddorn and Eatock 2008).

In the present paper, two and four truncated vertical cylinders in both sway and heave modes as a wave energy converter arrays will be used to predict the absorbed energy from ocean waves. The time dependent hydrodynamic radiation and exciting forces impulse response functions (which are used for the time marching of the equation of motion in order to find displacement, velocity, and acceleration of the wave energy converters) are predicted by the use of the transient free-surface wave Green function (Kara 2000, Kara 2010, Kara and Vassalos 2003, Kara and Vassalos 2005, Kara and Vassalos 2007, Kara 2011, Kara 2015, Kara 2016, Kara 2018). The present ITU-WAVE numerical results for two and four truncated vertical cylinder array systems that will be validated with analytical results (Kagemoto and Yue 1986). The effects of the separation distances and heading angles on relative capture width and interaction factor are then studied in order to determine the maximum absorbed power from ocean waves and the constructive and destructive effects.

2. Equation of Motion of Array Systems

A right-handed coordinate system is used to define the fluid action and a Cartesian coordinate system $\vec{x} = (x, y, z)$ is fixed to the body which is used for the solution of the linearized problem in the time domain. See Figure 1. Positive x-direction points forward, positive z-direction points upwards, and the $z = 0$ plane (or xy-plane) is coincident with calm water. The bodies undergo oscillatory motion about their

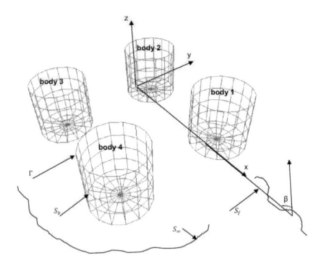

Figure 1. Coordinate system and surface of the wave energy converters.

mean positions due to incident wave field. The origin of the body-fixed coordinate system $\vec{x} = (x, y, z)$ is located at the centre of the xy plane. The solution domain consists of the fluid bounded by the free surface $S_f(t)$, the body surface $S_b(t)$, and the boundary surface at infinity S_∞. See Figure 1 (Kara 2000).

The following assumptions are taken into account in order to solve the physical problem. If the fluid is unbounded (except for the submerged portion of the body on the free surface), ideal (inviscid and incompressible), and its flow is irrotational (no fluid separation and lifting effect), the principle of mass conservation dictates the total disturbance velocity potential $\Phi(\vec{x}, t)$. This velocity potential is harmonic in the fluid domain and is governed by Laplace equation everywhere in the fluid domain as $\nabla^2 \Phi(\vec{x}, t) = 0$ and the disturbance flow velocity field $\vec{V}(\vec{x}, t)$ may then be described as the gradient of the potential $\Phi(\vec{x}, t)$ (e.g., $\vec{V}(\vec{x}, t) = \nabla\Phi(\vec{x}, t)$).

The dynamics of a floating body's unsteady oscillations are governed by a balance between the inertia of the floating body and the external forces acting upon it. This balance is complicated by the existence of radiated waves which results from the oscillations of the bodies and the scattering of the incident waves. This means that waves generated by the floating bodies at any given time will persist indefinitely and the waves of all frequencies will be generated on the free surface. These generated waves, in principle, affect the fluid pressure field and hence the body force of the floating bodies at all subsequent times. This situation introduces memory effects and is described mathematically by a convolution integral. Having assumed that the system is linear, the equation of motion of any floating bodies may be written in a form (Cummins 1962).

137

$$\sum_{k=1}^{6}(M_{kki} + a_{kki})\ddot{x}_{ki}(t) + (b_{kki} + B_{PTO-kki})\dot{x}_{ki}(t) + (C_{kki} + c_{kki} + C_{PTO-kki})\,x_{ki}(t) +$$
$$\int_{0}^{t} d\tau K_{kki}(t - \tau)\dot{x}_{ki}(\tau) = \int_{-\infty}^{\infty} d\tau K_{kDi}\,(t - \tau)\zeta(\tau) \tag{1}$$

where $i = 1,2,3,\ldots, N$ is the N number of body in the array systems. $k = 1,2,3,\ldots,6$ represent six-rigid body modes of surge, sway, heave, roll, pitch, and yaw respectively. The displacement of the floating bodies from its mean position in each of its rigid-body modes of motion is given as $x_k(t) = (1,2,3,\ldots,N)^T$, and the over-dots indicates differentiation with respect to time, $\ddot{x}_k(t)$ and $\dot{x}_k(t)$ are acceleration and velocity, respectively. M_{kk} is inertia matrix of the floating body and C_{kk} is linearized hydrostatic restoring force coefficients. As the same floating body is used in the array, the elements of both mass and restoring coefficients are equal to each other for each body $m_1 = m_2 = \cdots = m_N = m$ and $C_1 = C_2 = \cdots = C_N = C$, respectively. m and C are the mass and restoring coefficient for single body respectively.

$$M_{kk} = \begin{pmatrix} m_1 & \cdots & 0 \\ \vdots & \ddots & \vdots \\ 0 & \cdots & m_N \end{pmatrix} C_{kk} = \begin{pmatrix} C_1 & \cdots & 0 \\ \vdots & \ddots & \vdots \\ 0 & \cdots & C_N \end{pmatrix} \tag{2}$$

The radiation impulse response function $K_{kk}(t)$ is the force on the k-th body due to an impulsive velocity of the k-th body. The coefficients a_{kk}, b_{kk}, and c_{kk} account for the instantaneous forces proportional to the acceleration, velocity, and displacement respectively. The memory function $K_{kk}(t)$ accounts for the free surface effects which persist after the motion occurs. For the radiation problem, the term 'memory function' is used to distinguish this portion of the impulse-response function from the instantaneous force components outside of the convolution on the left-hand side of Equation (1). The coefficient a_{kk} is the time and frequency independent constant, it depends on the body geometry and is related to added mass. The coefficients b_{kk} and c_{kk} are the time and frequency independent constants and depend on the body geometry and forward speed and are related to damping and hydrostatic restoring coefficients respectively. The memory coefficient $K_{kk}(t)$ is the time dependent part and depends on body geometry, forward speed, and time. It contains the memory effect of the fluid response. The convolution integral on the left-hand side of Equation (1), whose kernel is a product of the radiation impulse response function $K_{kk}(t)$ and velocity of the floating body $\dot{x}_k(t)$, is a consequence of the radiated wave of the floating body. When this wave is generated, it affects the floating body at each successive time step (Oglivie 1964).

$$K_{kk}(t) = \begin{pmatrix} K_{11} & \cdots & K_{1N} \\ \vdots & \ddots & \vdots \\ K_{N1} & \cdots & K_{NN} \end{pmatrix}, a_{kk} = \begin{pmatrix} a_{11} & \cdots & a_{1N} \\ \vdots & \ddots & \vdots \\ a_{N1} & \cdots & a_{NN} \end{pmatrix}, b_{kk} = \begin{pmatrix} b_{11} & \cdots & b_{1N} \\ \vdots & \ddots & \vdots \\ b_{N1} & \cdots & b_{NN} \end{pmatrix},$$

$$c_{kk} = \begin{pmatrix} c_{11} & \cdots & c_{1N} \\ \vdots & \ddots & \vdots \\ c_{N1} & \cdots & c_{NN} \end{pmatrix}$$

(3)

The term $K_{kD}(t) = (K_{1D}, K_{2D}, K_{3D}, \ldots, K_{ND})^T$ on the right-hand side of Equation (1) are the components of the exciting force and moment's impulse response functions including Froude-Krylov and diffraction due to the incident wave elevation $\zeta(t)$ which is the arbitrary wave elevation and defined at the origin of the coordinate system (Figure 1) in the body-fixed coordinate system. The kernel $K_{kD}(t)$ is the diffraction impulse response function; the force on the k-th body due to uni-directional impulsive wave elevation with a heading angle of β (King 1987).

B_{PTO-kk} is time independent and frequency dependent Power-Take-Off (PTO) damping coefficient matrix, whilst C_{PTO-kk} is time and frequency independent restoring coefficient matrices for each mode of motion. The diagonal elements of B_{PTO-kk} is taken as the damping coefficient of the single body at natural frequency of each mode $B_{PTO} = B_{single}(\omega_n)$ in order to absorb maximum power (Budal and Falnes 1976) while the off-diagonal terms are considered to be zero for simplicity. The elements of C_{PTO-kk} is considered to be zero for heave mode while for the sway mode, the diagonal elements of C_{PTO-kk} are taken as hydrostatic restoring coefficient of heave mode in order to have the same natural frequency and displacement in both heave and sway modes. In this case, it would be possible to compare heave and sway motions and power variables directly in order to decide which mode of motion is more beneficial for power absorption.

$$B_{PTO-kk} = \begin{pmatrix} B_{single}(\omega_n) & \cdots & 0 \\ \vdots & \ddots & \vdots \\ 0 & \cdots & B_{single}(\omega_n) \end{pmatrix} C_{PTO-kk} = \begin{pmatrix} C_1 & \cdots & 0 \\ \vdots & \ddots & \vdots \\ 0 & \cdots & C_N \end{pmatrix}$$

(4)

Once the restoring matrix, inertia matrix, and fluid forces, e.g., radiation and diffraction forces are known; the equation of motion of floating body Equation (1) may be time marched using the fourth-order Runge-Kutta method (Kara 2000, Kara 2010, Kara and Vassalos 2003, Kara and Vassalos 2005, Kara and Vassalos 2007, Kara 2011, Kara 2015, Kara 2016, Kara 2018).

3. ITU-WAVE

The hydrodynamics functions in the present paper are predicted with in-house ITU-WAVE three-dimensional direct time domain numerical code. ITU-WAVE transient wave-structure interaction code which is coded using C++ was validated against experimental, analytical, and other published numerical results (Kara 2000, Kara 2010, Kara and Vassalos 2003, Kara and Vassalos 2005, Kara and Vassalos 2007, Kara 2011, Kara 2015, Kara 2016, Kara 2016) and used to predict the sea-keeping characteristics (e.g., radiation and diffraction), motions, resistance, added-resistance, hydroelasticity of the floating bodies, wave power absorption from ocean waves with latching control.

The fluid boundaries are described by the use of Boundary Integral Equation Methods (BIEM) with Neumann-Kelvin linearization in the context of potential theory. The exact initial boundary value problem is then linearized using the free stream as a basis flow and replaced by the boundary integral equation applying Green theorem over three-dimensional transient free surface Green function. The resultant boundary integral equation is discretized using quadrilateral elements over which the value of the potential is assumed to be constant and solved using the trapezoidal rule to integrate the memory part of the transient free surface Green function in time. The free surface and body boundary conditions are linearized on the discretized collocation points over each quadrilateral element in order to obtain algebraic equation.

4. Comparison with Analytical Results

The present ITU-WAVE numerical results are compared with the analytical results of two and four truncated vertical cylinders (Kagemoto and Yue 1986) in order to validate the present numerical predictions.

4.1 Two Truncated Vertical Cylinder Arrays

Two truncated vertical cylinders in Figure 2 are used for numerical analysis as a first test case. It is assumed that the two cylinders have the same draught and radius R, although the present method can be applied for different draughts and radii. The truncated cylinders have the radius of R, draught of 2R and hull separation to diameter ratio of d/D = 1.3. It is assumed that two truncated cylinders are free for sway and heave modes and fixed for other modes. These two truncated cylinders are studied to predict sway and heave radiation and diffraction impulse response functions in time and added-mass, damping coefficients, and exciting forces in frequency domain. The time domain and frequency domain results are related to each other through Fourier transforms in the context of linear analysis. The

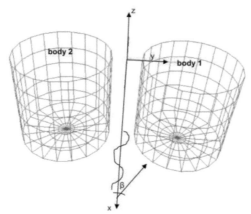

Figure 2. Two truncated vertical cylinders with the hull separation to diameter d/D = 1.3.

present ITU-WAVE numerical results for sway and heave added-mass and damping coefficients and exciting forces (which are the sum of the diffraction and Froude-Krylov forces) with heading angle $\beta = 90°$ are compared with the analytical results of (Kagemoto and Yue 1986).

Figure 3 shows the convergence test of radiation and diffraction Impulse Response Functions (IRFs) for sway and heave modes. As two truncated vertical

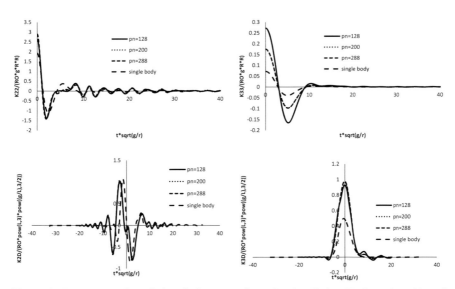

Figure 3. Two truncated vertical cylinders—non-dimensional radiation $K_{22}(t)$ and $K_{33}(t)$ and diffraction sway and heave $K_{2D}(t)$ and $K_{3D}(t)$ IRFs at d/D = 1.3 and beam seas $\beta = 90°$.

cylinders are symmetric in terms of xz-coordinate plane of the reference coordinate system, only single hull form is discretized for numerical analysis. Numerical experience showed that numerical results are not very sensitive in terms of non-dimensional time step size $t * sqrt(g/R)$ (where t is time, g gravitational acceleration, R radius) of 0.01, 0.03, and 0.05 over the range of panel numbers of 128, 200, and 288 on single body of two truncated vertical cylinders, whilst the numerical results are quite sensitive in terms of panel numbers as can be seen in Figure 3 and the results at panel number 200 on single hull form is converged and used for the present ITU-WAVE numerical calculations for both two and single truncated vertical cylinders with the non-dimensional time step size of 0.05.

It may be noticed that the magnitude of radiation IRFs of two cylinders in heave mode (Figure 3) is approximately twice of IRF of single cylinder, while it is less than double in the case of sway mode. The other distinctive difference of IRF of single and two cylinders in Figure 3 is the behavior of these IRFs functions in longer times in sway mode. IRFs of two cylinders have oscillations over longer times with decreasing amplitude in sway mode while single cylinder IRF decays to zero just after first oscillation. This behavior of IRF implicitly means that the energy between two cylinders is trapped in the gap and only a minor part of the energy is radiated outwards each time. The wave is reflected off the hull while all energy is dissipated in the case of single cylinder in sway mode and in both two and single cylinders in heave mode. It is expected that geometry of the two bodies would significantly affect the radiated, diffracted, and trapped waves which result due to standing waves in the gap. In the case of diffraction IRF in Figure 3, there are no significant differences in sway mode between single and two cylinders' IRFs except slight shift, whilst it is doubled in heave mode.

The time dependent radiation and exciting IRFs in time domain are related to the frequency dependent added-mass and damping cocfficients and force amplitude, respectively in frequency domain through Fourier transforms when the motion is considered to be a time harmonic motion. Added-mass $A_{22}(\omega)$ and $A_{33}(\omega)$, damping coefficients $B_{22}(\omega)$ in Figure 4 and exciting forces amplitudes $F_2(\omega)$ and $F_3(\omega)$ in Figure 5 are obtained by the Fourier transform of radiation sway IRF $K_{22}(t)$ and radiation heave IRF $K_{33}(t)$ of Figure 3, and diffraction sway IRF $K_{2D}(t)$ and diffraction heave IRF $K_{3D}(t)$ of Figure 3, respectively.

ITU-WAVE numerical results of added-mass and damping coefficients in sway mode and added-mass coefficients in heave mode of two cylinders are in satisfactory agreement with the analytical prediction of (Kagemoto and Yue 1986) as can be seen in Figure 4. In addition, the added-mass and damping coefficients of the two cylinders array are presented in Figure 4 and compared to those from the single cylinders. The behaviours of two cylinders in sway mode in Fig. 4 are significantly different from those of single cylinder due to trapped waves and

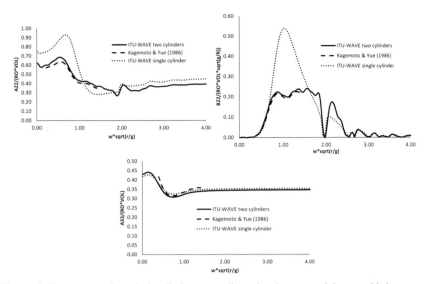

Figure 4. Two truncated vertical cylinders—non-dimensional sway and heave added-mass and damping coefficients at d/D = 1.3.

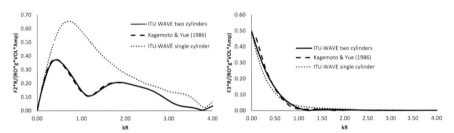

Figure 5. Two truncated vertical cylinders—non-dimensional sway and heave exciting force amplitude at d/D = 1.3 and beam seas $\beta = 90°$.

hydrodynamic interactions in the gap of two cylinders. However, single and two cylinders added-mass result in heave mode do not show such difference as most of the energy is dissipated in this mode.

The effects of diffraction hydrodynamic interactions in sway mode (at which interactions are effective in the whole frequency range) are much stronger than in heave mode as can be observed in Figure 5. The effect of this interaction in sway mode is even stronger in a limited frequency range which is of interest to the motions of the bodies in array systems and are around kR = 0.5 and kR = 2.0 of non-dimensional frequency in radiation and diffraction sway mode in Figure 4 and Figure 5 respectively.

4.2 Four Truncated Vertical Cylinder Arrays

Four truncated vertical cylinders in Figure 6 are used for numerical analysis as the second test case. As in two cylinders, it is assumed that four cylinders have the same draught and radius. Four truncated cylinders have the radius of R and draught of 2R and hull separation to diameter ratio of d/D = 2.0. It is assumed that four truncated cylinders are free for sway mode and fixed for other modes and are studied to predict sway radiation and diffraction IRFs in time and added-mass, damping coefficients, and exciting force amplitude in frequency domain. The present ITU-WAVE numerical results for sway added-mass and damping coefficients and exciting force amplitude with heading angle $\beta = 90°$ are compared with the analytical results of (Kagemoto and Yue 1986).

Figure 6 shows the convergence test of radiation and diffraction IRFs for sway mode. As four truncated vertical cylinders are symmetric, only single hull form is discretized for numerical analysis. Numerical experience showed that numerical results at panel number 200 on single hull form are converged and used for the present ITU-WAVE numerical calculations for both four and single truncated vertical cylinders with the non-dimensional time step size of 0.05.

When two (Figure 3) and four (Figure 6) truncated vertical cylinders' radiation IRFs are compared, it can be observed that the amplitude of radiation IRFs of four truncated cylinders are approximately 2.5 times bigger than two cylinders' radiation IRFs and four cylinders' IRFs have oscillations over longer times with decreasing amplitude in sway mode compared to two cylinders' IRFs. This behaviour implicitly means that more energy captured between bodies in four cylinders than two cylinders. The same outcome is valid for diffraction IRFs too.

Figure 7 shows added-mass $A_{22}(\omega)$, damping coefficients $B_{22}(\omega)$, and exciting forces amplitudes $F_2(\omega)$ which are obtained by the Fourier transform of radiation sway IRF $K_{22}(t)$ of Figure 6, and diffraction sway IRF $K_{2D}(t)$ of Figure 6, respectively. ITU-WAVE numerical results of four cylinders are satisfactory agreements with those of (Kagemoto and Yue 1986) as can be seen in Figure 7.

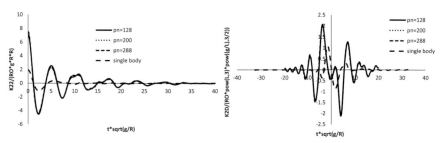

Figure 6. Four truncated vertical cylinders—non-dimensional radiation $K_{22}(t)$ and diffraction $K_{2D}(t)$ sway IRFs at d/D = 2.0 and beam seas $\beta = 90°$.

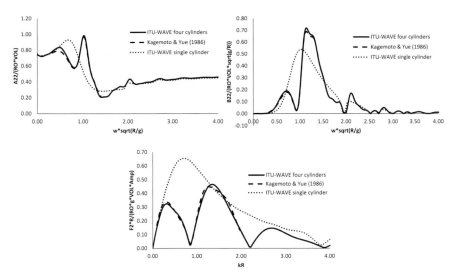

Figure 7. Four truncated vertical cylinders—non-dimensional sway added-mass, damping, and exciting force amplitude at d/D = 2.0 and beam seas $\beta = 90°$.

There would not be energy transfer or radiated waves from floating body to sea when the damping coefficients are zero as can be observed in Figure 7. It may be noticed there are three resonances behaviors in damping coefficients in sway mode which implies that high standing waves occur between the maximum and minimum damping coefficients (Ohkusu 1969, van Oortmerssen 1979). It may be noticed the peaks are finite at non-dimensional resonance frequencies as some of the wave energy dissipate under the floating body and radiates to the far field.

5. The Interactions of Bodies on Array System

The radiation IRFs for sway and heave modes in the case of two interacting bodies for each body in arrays are presented for the range of different separation distances in Figure 8. The radiation IRFs $K_{12}(t)$ which represents the interactions between two truncated vertical cylinders is very strong and the same order with $K_{11}(t)$, whilst the interactions become weaker as the separation distance between interacting bodies is increased. The interactions IRFs on body 1 and body 2 have the same magnitude and sign as it is presented in Figure 8. Giving one body an impulsive velocity in one mode causes a force in the same mode on the other body after some finite time t, which is the time it takes the wave to move the distance between bodies. This means that energy is trapped in the gap between bodies and only a minor part of the energy is radiated outwards each time the wave is reflected off the body.

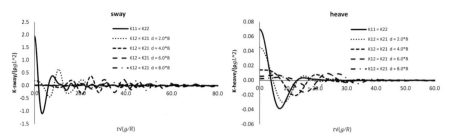

Figure 8. Two truncated vertical cylinders—non-dimensional radiation sway and heave IRFs at a range of different separation distances.

It may be noticed from Figure 8 that the dominant part of the interactions between these two vertical cylinders are shifted to the larger times as the separation distance increases. The exciting force IRFs (which is the sum of diffraction and Froude-Krylov forces) for sway and heave modes for each body in arrays is presented in Figure 9 for the range of different separation distances at heading angle $\beta = 90°$.

It may be noticed from Figure 9 that when the separation distance increases, the interaction between incident wave and the first body which interacts with the incident wave first is delayed for longer times, whilst it is contrary for the second body which is in the wake of the first body in the case of heading angle $\beta = 90°$.

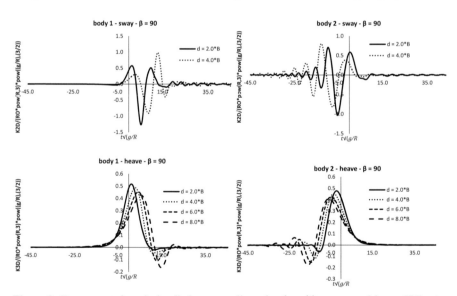

Figure 9. Two truncated vertical cylinders—non-dimensional exciting sway and heave IRFs at a range of different separation distances and beam seas $\beta = 90°$.

Radiation IRFs of sway and heave modes for four truncated vertical cylinders for each body in arrays are presented in Figure 10 for the separation distance of 2.0*B and 8.0*B in order to predict the effects of the interaction of bodies on IRFs in arrays. As in two truncated vertical cylinders, in the case of small separation distance of 2.0*B, the interaction between the bodies is quite significant while the interaction's effects decrease for larger separation distance of 8.0*B and the maximum response amplitudes are shifted to larger times.

Exciting sway and heave IRFs for four truncated cylinder for each body in arrays at beam seas $\beta = 90°$ and separation distance of d = 2.0*B are presented in Figure 11. It may be noticed in Figure 11 that the exciting forces on body 1 and

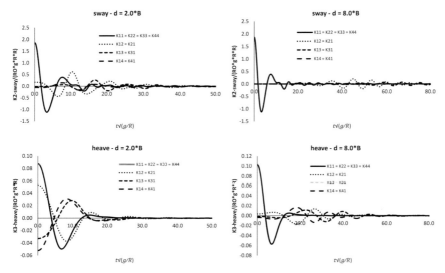

Figure 10. Four truncated vertical cylinders—non-dimensional radiation sway and heave IRF sat d = 2.0*B and d = 8.0*B separation distances.

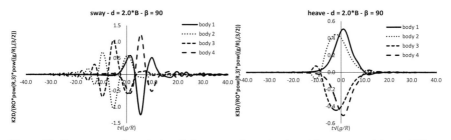

Figure 11. Four truncated vertical cylinders—non-dimensional exciting sway and heave IRFs at separation distance d = 2.0*B and beam seas $\beta = 90°$.

body 4 which are on the wake of body 1 as well as on body 2, and body 3 which is on the wake of body 2 (Figure 11) have the same magnitudes but opposite signs. This is the numerical prediction that is expected as the truncated four cylinders are symmetric both in x- and y-directions. The wave is reflected off the bodies and translating across the gap between the hulls to hit the other body. The detailed discussion of how the distance between the devices in an array affects the capture efficiency can be found in (Nader et al. 2012).

6. Response Amplitude Operators (RAOs) of Sway and Heave Motions

As sway mode does not have restoring coefficient due to motion on the sea, PTO restoring coefficient of the truncated circular cylinders are taken as hydrostatic restoring coefficient of heave mode in order to have the same natural frequency and displacement in both heave and sway modes as mentioned in Section 2. RAOs of two and four truncated vertical cylinders in sway and heave modes at separation distance d = 2.0*B and beam seas β = 90° are presented in Figure 12. RAOs are obtained by the use of Equation (1) after time simulation of equation of motion, i.e., Equation (1) achieving the steady state condition at the range of the different frequencies.

It may be noticed that the motion of the second body which is at the wake of the first body at beam seas β = 90° in the case of two bodies (Figure 12) and that of the second and third bodies which are at the wake of the first and fourth bodies, respectively, in the case of four truncated cylinder (Figure 11) is higher around resonance frequency in both sway and heave modes. This is mainly due to the trapped waves between bodies in arrays. In the case of four truncated vertical cylinders in Figure 11, it may also be noticed that the motion of the first and fourth body in the array system and the second and third body is exactly the same which is due to symmetry in terms of x- and y-coordinate systems (Figure 11).

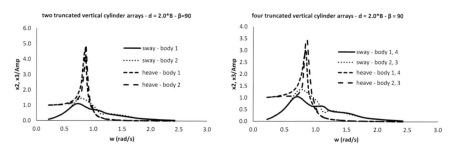

Figure 12. Two (left) and four (right) truncated vertical cylinders—sway and heave motions RAOs at separation distance d = 2.0*B and beam seas β = 90°.

The heave motion amplitude is more than two times of sway motion amplitude around resonance frequency. This is mainly due to the damping coefficients as sway mode damping coefficients are much bigger than heave damping coefficients around resonance frequency. As the mass of the bodies in arrays is balanced by the restoring forces, the motions of each body in arrays are controlled by the damping coefficients around the resonance frequency. Since sway mode have much bigger damping coefficients, sway motion is damped around resonance frequency compared to heave motion. However, the sway motion distributed wider frequency range as heave motion mainly concentrated at resonance frequency.

7. Instantaneous and Mean Absorbed Power

The instantaneous power $P_{ins_{k_i}}(t)$ absorbed by Power-Take-Off (PTO) system for each mode and body in arrays is directly proportional to exciting (which is the sum of diffraction and Froude-Krylov forces) and radiation forces on bodies in arrays and is defined as

$$P_{ins_{k_i}}(t) = [F_{exc_{k_i}}(t) + F_{rad_{k_i}}(t)] \cdot \dot{x}_{k_i}(t) \tag{5}$$

where $F_{exc_i}(t)$ = exciting forces which are due to incident and diffracted waves, $F_{rad_i}(t)$ = radiation forces which are due to the oscillation of bodies in arrays. $F_{exc_i}(t)$ and $F_{rad_i}(t)$ are given for each body and mode as in Equation (1) (Kara 2000, Kara 2010, Kara and Vassalos 2003, Kara and Vassalos 2005, Kara and Vassalos 2007, Kara 2011, Kara 2015, Kara 2016, Kara 2016)

$$F_{exc_{k_i}}(t) = F_{k_i}(t) = \int_{-\infty}^{\infty} d\tau K_{kD_i}(t - \tau)\, \zeta(\tau) \tag{6}$$

$$F_{rad_{k_i}}(t) = F_{kk_i}(t) = -a_{kk_i}\ddot{x}_{k_i}(t) - b_{kk_i}\dot{x}_{k_i}(t) - c_{kk_i}x_{k_i}(t) - \int_0^t d\tau K_{kk_i}(t-\tau)\,\dot{x}_{k_i}(\tau) \tag{7}$$

The power due to exciting forces $P_{exc_{k_i}}(t) = F_{exc_{k_i}}(t) \cdot \dot{x}_{k_i}(t)$ is the total absorbed power from the incident and diffracted waves, whilst the power due to radiation forces $P_{rad_{k_i}}(t) = F_{rad_{k_i}}(t) \cdot \dot{x}_{k_i}(t)$ is the power radiated back to sea due to the oscillation of bodies on arrays. The mean (average) power $\bar{P}_{ins_{k_i}}(t)$ absorbed by the PTO system over a time range T is given by

$$\bar{P}_{ins_{k_i}}(t) = \frac{1}{T}\int_0^T dt \cdot [F_{exc_{k_i}}(t) + F_{rad_{k_i}}(t)] \cdot \dot{x}_{k_i}(t) \tag{8}$$

The averaging time T must be much larger than the characteristics period of the incident wave which is approximately from 5 s to 15 s. In order to avoid the transient effects, only the last half of the time domain results are taken into account for the prediction of the mean absorbed power using Equation (8) and other time dependent parameters. The total mean absorbed power $\overline{P}_{T_k}(t)$ for N number of array systems and for mode k is given as

$$\overline{P}_{T_k}(t) = \sum_{i=1}^{N} \overline{P}_{insk_i}(t) \qquad (9)$$

The total absorbed power with two and four truncated vertical cylinders in sway and heave modes at heading angle $\beta = 90°$ and separation distance d = 4.0*B are presented in Figure 13 using asymptotic value of Equation (9) for a range of incident waves.

The power absorption is concentrated around the resonance frequency in heave mode, whilst power absorption has wider frequency range and the frequency bandwidth of power absorption is much larger in sway mode than in heave mode. It may be noticed that power absorption is doubled in sway mode as the number of bodies in arrays is increased from two to four, whilst absorbed power is not changed significantly in the case of heave mode. As mentioned above, 200 panels on a single body with 0.05 non-dimensional time step and 24 wave frequency, ranging from 0.222 rad/s to 2.43 rad/s, used as the numerical results are converged at this panel number and non-dimensional time step. The computational time is 33 minutes for two truncated vertical cylinders, while it is 41 minutes for four truncated vertical cylinders.

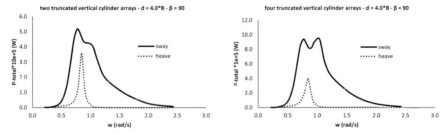

Figure 13. Two (left) and four (right) truncated vertical cylinders—sway and heave total absorbed power at separation distance d = 4.0*B and beam seas $\beta = 90°$.

8. Capture Width and Relative Capture Width

A good wave energy converter is a converter which absorbs as much energy as possible from the incident wave. However, to absorb maximum power, the body should interact with the sea significantly (i.e., the characteristic frequencies of the incident waves should be close to the resonance frequency of the body). The quantity employed to evaluate the performance of a device in terms of power absorption is the capture width. At a given frequency, it is the ratio between the total mean power absorbed and the mean power per unit crest wave width of the incident wave train. The capture width $l_{cwk_i}(\omega)$ for each incident wave frequency ω, for each mode k, and for a body i in array systems is defined as (Budal and Falnes 1976)

$$l_{cwk_i}(\omega) = \frac{\overline{P}_{insk_i}(\omega)}{P_w(\omega)} \tag{10}$$

where $\overline{P}_{insk_i}(\omega)$ is the mean absorbed power and is obtained as the asymptotic value of the mean of the instantaneous power Equation (8), $P_w(\omega) = \rho g^2 \zeta_0^2 / 4\omega$ is the wave power in the incident wave train per unit crest length, ζ_0 being the incident wave amplitude. Relative capture width is obtained by dividing capture width with width B of bodies in arrays

$$l_{rcwk_i}(\omega) = l_{cwk_i}(\omega)/B \tag{11}$$

If the relative capture width is greater than 1 at any incident wave frequencies, this implies that the absorbed power can be greater than the incident wave power at these frequencies. The detailed discussion about capture width can be found in (Evans and Porter 1996).

8.1 Separation Distance Effect on Capture Width

The relative capture width of two truncated vertical cylinders at heading angle $\beta = 90°$ in sway and heave mode are presented in Figure 14 using Equation (11).

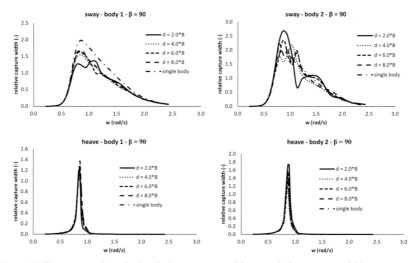

Figure 14. Two truncated vertical cylinders—sway and heave relative capture width at a range of separation distances and beam seas $\beta = 90°$.

Relative capture width in both sway and heave modes for body 1 is less than single body at all frequency range, whilst for body 2 it is contrary. This implicitly means that the wave interaction with body 1 has the destructive effects, while the wave interaction has constructive effects for body 2 as wave energy is trapped between bodies in arrays. This implies that the absorbed power can be greater than the incident wave power at these frequencies for body 2.

Due to symmetry in both x- and y-directions, in the case of four truncated vertical cylinders, capture width of body 1 and body 4 is the same as well as of body 2 and body 3 in Figure 15. The capture width for four bodies in beam seas $\beta = 90°$ using Equation (11) is presented in Figure 15.

As in the case of two truncated cylinders in sway mode, the capture width for body 1 and body 4 has destructive effects (e.g., absorbed less power than isolated body) around resonance frequency at which most power is captured, whilst body 2 and body 3 in the array system has constructive effects. It may be noticed in Figure 15 that the power is absorbed in wider frequency ranges in sway mode, whilst the power is absorbed only around resonance frequency in the case of heave mode. It may also be noticed that in heave mode, wave interactions have destructive effects compared to single body apart from separation distance of d = 6.0*B for body 2 and body 3.

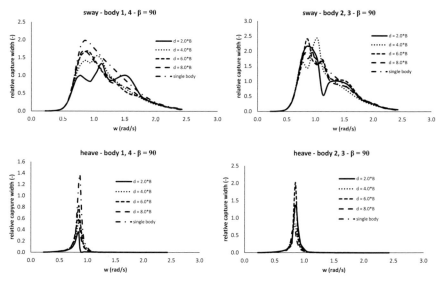

Figure 15. Four truncated vertical cylinders—sway and heave relative capture width at a range of separation distances and beam seas $\beta = 90°$.

8.2 Heading Angle Effect on Capture Width

The effect of heading angles on capture width at separation distance d = 2.0*B in both sway and heave modes are presented in Figure 16 which shows the maximum power absorbed in the heading angle of $\beta = 90°$ in both sway and heave modes. When compared with heave mode which only absorbed power around resonance frequency, the bandwidth of absorbed power is larger in sway mode.

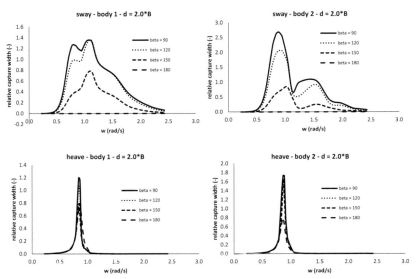

Figure 16. Two truncated vertical cylinders—sway and heave relative capture width at separation distance d = 2.0*B and a range of different heading angles.

The effect of heading angles from 90 degrees to 180 degrees for four truncated vertical cylinders in both sway and heave modes are presented in Figure 17 which shows relative capture width for separation distance d = 2.0*B.

More power is absorbed from ocean waves in heading angle $\beta = 120°$ than other heading angle in sway mode for body 1 and body 2, whilst for body 2 and body 3 almost the same level of power is absorbed at heading angle $\beta = 120°$ and $\beta = 90°$ in Figure 17. In the case of heave mode, more power is absorbed in heading angle $\beta = 180°$ for body 1 and body 4, whilst the heading angle $\beta = 120°$ is the heading angle at which more power is absorbed for body 2 and body 3 in Figure 17.

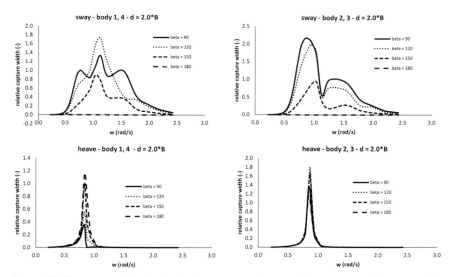

Figure 17. Four truncated vertical cylinders—sway and heave relative capture width at separation distance d = 2.0*B and a range of different heading angles.

9. Interaction Factor

The interaction factor due to diffracted and radiated waves gives information about the mean interactions between bodies of array systems and represents the mean gain factor for each body of the interacting systems of N bodies. Two kinds of interaction factors are used in the present paper. First one is the standard $q_{k_i}(\omega)$ for a given incident wave frequency, mode k and body i in the array system. Second one is the modified $q_{mod_{k_i}}(\omega)$. The overall power production from bodies in arrays are very sensitive to the interaction factor. Depending on separation distances between bodies, geometry, control strategies, wave length, and heading angles, standard $q_{k_i}(\omega)$ factor can have constructive $(q_{k_i}(\omega) > 1)$ or destructive $(q_{k_i}(\omega) < 1)$ effect and given as (Thomas and Evans 1981).

$$q_{k_i}(\omega) = \frac{\overline{P}_{ins_{k_i}}(\omega)}{\overline{P}_{ins_{k_0}}(\omega)} \tag{12}$$

where $\overline{P}_{ins_{k_i}}(\omega)$ is the absorbed power from a single truncated vertical cylinder. The constructive $(q_{k_i}(\omega) > 1)$ effect implicitly means that power absorption from array system increases in that particular wave frequency, whilst the power absorption decreases in the case of destructive $(q_{k_i}(\omega) < 1)$ effect.

In the case of modified interaction factor $q_{modk_i}(\omega)$ which is given in Equation (13), the dominant wave interaction which results in maximum power absorption in the array systems and occurs around natural frequency of WEC is taken into account and weaker wave interaction in which the power absorption is lower are filtered out from the power prediction (Babarit 2010).

$$q_{modk_i}(\omega) = \frac{\bar{P}_{ins_{k_i}}(\omega) - \bar{P}_{ins_{k_0}}(\omega)}{\max_\omega \bar{P}_{ins_{k_0}}(\omega)} \tag{13}$$

The modified interaction factor $q_{modk_i}(\omega)$ in Equation (13) can have constructive $(q_{k_i}(\omega) > 0)$ or destructive $(q_{k_i}(\omega) < 0)$ effect. The detailed discussion about the q–factors of a finite array of wave power devices can be found in (Mavrakos 1997).

9.1 Separation Distance Effect on Interaction Factor

The interaction factor $q_{k_i}(\omega)$ for sway and heave modes at heading angle $\beta = 90°$ and different range of separation distances is presented in Figure 18 using Equation (11). It can be seen in Figure 17 that in both sway and heave modes for body 1, the separation distance has destructive effects.

As expected, the strongest effects are due to shortest separation distance d = 2.0*B. However, for body 2 which is on the wake of the body 1 (Figure 2), wave interactions have both constructive and destructive effects depending on incident wave frequencies.

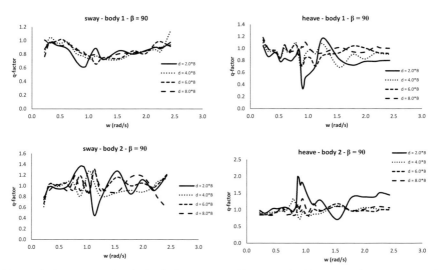

Figure 18. Two truncated vertical cylinders—sway and heave standard interaction factor at a range of different separation distances and heading angle $\beta = 90°$.

The modified interaction factor $q_{modki}(\omega)$ for two truncated vertical cylinders in both sway and heave modes are presented in Figure 19 using Equation (12). The effects of incident wave frequencies away from the resonance frequency (e.g., in very low and very high frequency regions) are filtered out in Figure 19. This is valid for both sway and heave modes.

It may be noticed in that in Figure 19, for body 1 in sway and heave modes the wave interactions are destructive, while for body 2 the wave interactions are both constructive and destructive depending on the wave frequency. As in the relative capture width in Figure 14–17, destructive or constructive wave interaction effects have wider frequency range in sway mode as compared to heave mode at which power absorption is mainly around resonance frequency as can be observed from Figure 19.

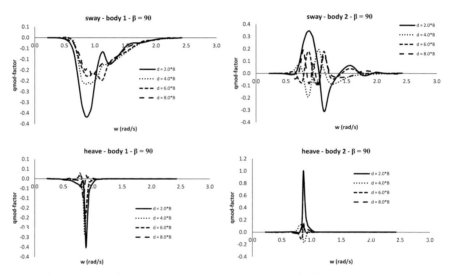

Figure 19. Two truncated vertical cylinders—sway and heave modified interaction factor at a range of different separation distances and heading angle $\beta = 90°$.

9.2 Heading Angles Effect on Interaction Factor

The modified interaction factor $q_{modki}(\omega)$ Equation (13) for two truncated vertical cylinders at separation distance d = 2.0*B and different heading angles in both sway and heave modes are presented in Figure 19 which shows that heading angle $\beta = 150°$ in sway mode for both body 1 and body 2 has most favourable constructive effects at different incident wave frequencies.

In case of heave mode at all heading angles, body 1 has destructive effect as can be seen in Figure 20 while at heading angle $\beta = 90°$, body 2 has maximum constructive effect as compared to other heading angles.

The modified interaction factor $q_{mod_{ki}}$ (ω) in sway mode for four truncated vertical cylinders is presented in Figure 21 for the separation distance d = 2.0*B

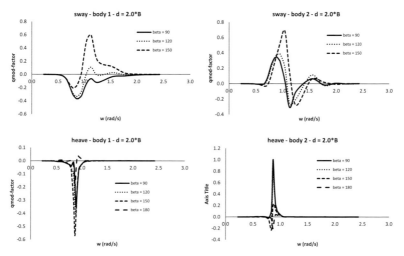

Figure 20. Two truncated vertical cylinders—sway and heave modified interaction factor $q_{mod_{ki}}$ (ω) at a range of different heading angles and separation distance d = 2.0*B.

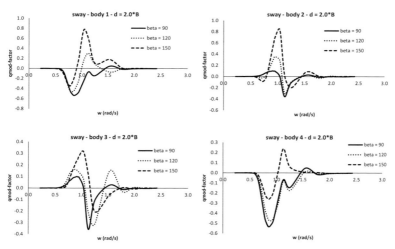

Figure 21. Four truncated vertical cylinders—sway modified interaction factor $q_{mod_{ki}}$ (ω) at a range of different heading angles and separation distance d = 2.0*B.

and different heading angles. As in the case of two cylinders Figure 20 in sway mode, the most favourable heading angle is $\beta = 150°$ which has more constructive effects as compared to other heading angles for all bodies in the array system.

The modified interaction factor $q_{mod_{kj}}(\omega)$ in heave mode for four truncated vertical cylinders is presented in Figure 22 for the separation distance d = 2.0*B and different heading angles.

Body 1, body 3 (except heading angle $\beta = 90°$), and body 4 in all heading angles has destructive effects in heave mode in the case of four truncated vertical cylinders as can be seen in Figure 21, whilst body 2 in the array system has constructive effects in all heading angles except heading angle $\beta = 180°$.

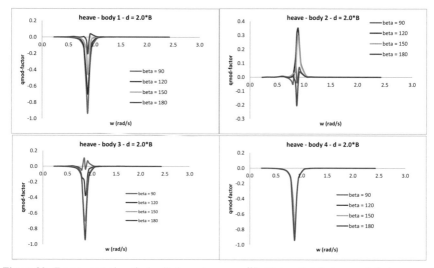

Figure 22. Four truncated vertical cylinders—heave modified interaction factor $q_{mod_{kl}}(\omega)$ at a range of different heading angles and separation distance d = 2.0*B.

10. Conclusion and Outlook

The numerical capability of present ITU-WAVE three-dimensional transient wave-structure interaction panel method is extended to predict the power absorption from ocean waves in array systems. The present numerical results in both sway and heave modes are validated with analytical array results after obtaining the added-mass and damping coefficients as well as exciting force amplitude using Fourier transforms of radiation and diffraction IRFs, respectively, in order to present the results in frequency domain.

The numerical experience shows that power absorptions in sway modes in any separation distances and heading angles are much higher than in heave mode and have wider bandwidth in frequency range in sway mode. The modified interaction factor, the q-factor, has more constructive effect in heading angle $\beta = 150°$ than any other heading angles in the case of sway mode in both two and four truncated vertical cylinders.

In case of heave mode modified interaction factor, the q-factor, the most effective heading angle for constructive effect for two and four truncated cylinders are mixed. The maximum constructive effect is heading angle $\beta = 90°$ for two truncated cylinders, whilst it is heading angle $\beta = 120°$ for body 2 in the array system and is heading angle $\beta = 90°$ for body 3 in the case of four truncated vertical cylinders.

The numerical experience also shows that if the bodies in arrays are close, the wave interactions are stronger at any wave headings and separation distances. These interaction effects' are significantly diminished and shifted to larger times when the separation distances are increased. The wave interaction in radiation heave mode is much stronger than in radiation sway mode in both two and four truncated vertical cylinders.

References

ANSYSA, W. (2013). Inc., AQWA manual Release 15.0, 2013.

Babarit, A. (2010). Impact of long separating distances on the energy production of two interacting wave energy converters. Ocean Engineering, 37(8-9): 718–729.

Babarit, A. (2013). On the park effect in arrays of oscillating wave energy converters. Renewable Energy, 58: 68–78.

Babarit, A., Borgarino, B., Ferrant, P. and Clement, A. (2010). Assessment of the influence of the distance between two wave energy converters on energy production. IE Trenewable power generation, 4: 592–601.

Bellew, S., Stallard, T. and Stansby, P. (2009). Optimisation of a heterogeneous array of heaving bodies. Proceedings of the 8th European Wave and Tidal Energy Conference (EWTEC 2009), 1–9.

Bertram, V. (1990). Ship motions by rankine source method. Ship Technology Research, 37(4): 143–152.

Borgarino, B., Babarit, A. and Ferrant, P. (2012). Impact of wave interactions effects on energy absorption in large arrays of wave energy converters. Ocean Engineering, 41: 79–88.

Breit, S. and Sclavounos, P. (1986). Wave interaction between adjacent slender bodies. Journal of Fluid Mechanics, 165: 273–296.

Budal, K. and Falnes, J. (1976). Optimum operation of wave power converter. Internal Report, Norwegian University of Science and Technology.

Budal, K. (1977). Theory for absorption of wave power by a system of interacting bodies. Journal of Ship Research, 21(4): 248–253.

Child, B. and Venugopal, V. (2010). Optimal configurations of wave energy device arrays. Ocean Engineering, 37: 1402–1417.

Child, B.,Cruz, J. and Livingstone, M. (2011). The development of a tool for optimising arrays of wave energy converters. Proceedings of the Ninth European Wave and Tidal Energy Conference, EWTEC.

Child, B.F.M. and Venugopal, V. (2007). Interaction of waves with an array of floating wave energy devices. Proceedings of the 7th European Wave and Tidal Energy Conference (EWTEC 2007).

Cummins, W.E. (1962). The Impulse response function and ship motions. Shiffstechnik, 9: 101–109.

De Backer, G., Vantorre, M., Beels, C., De Rouck, J. and Frigard, P. (2009). Performance of closely spaced point absorbers with constrained floater motion. Proceedings of the 8th European Wave and Tidal Energy Conference (EWTEC 2009), 806–817.

Delhommeau, G. (1993). Seakeeping codes aquadyn and aquaplus. *In*: Proc. of the 19th WEGEMT School, Numerical Simulation of Hydrodynamics: Ships and Offshore Structures.

Evans, D.V. and Porter, R. (1996). Efficient calculation of hydrodynamic properties of OWC type devices. *In*: OMAE – Volume I – Part B; 123–132.

Fitzgerald, C. and Thomas, G.P. (2007). A preliminary study on the optimal formation of an array of wave power devices. Proceedings of the 7th European Wave and Tidal Energy Conference.

Folley, M., Babarit, A., Child, B., Forehand, D., O'Boyle, L., Silverthorne, K., Spinneken, J., Stratigaki, V. and Troch, P. (2012). A review of numerical modelling of wave energy converter arrays. ASME201231st International Conference on Ocean, Offshore and Arctic Engineering, 2012. American Society of Mechanical Engineers, 535–545.

Justino, P.A.P. and Clement, A.H. (2003). Hydrodynamic performance for small arrays of submerged spheres. Proceedings of the 5th European Wave and Tidal Energy Conference (EWTEC 2003), 266–273.

Kagemoto, H. and Yue, D.K.P. (1993). Hydrodynamic interaction analyses of very large floating structures. Marine Structures, 6: 295–322.

Kagemoto, H. and Yue, D.K.P. (1986). Interactions among multiple three-dimensional bodies in water waves: an exact algebraic method. Journal of Fluid Mechanics, 166: 189–209.

Kara, F. (2000). Time domain hydrodynamics and hydroelastics analysis of floating bodies with forward speed. PhD thesis, University of Strathclyde, Glasgow, UK.

Kara, F. and Vassalos, D. (2003). Time domain prediction of steady and unsteady marine hydrodynamic problem. International Shipbuilding Progress, 50(4): 317–332.

Kara, F. and Vassalos, D. (2005). Time domain computation of wave making resistance of ships. Journal of Ship Research, 49(2): 144–158.

Kara, F. and Vassalos, D. (2007). Hydroelastic analysis of cantilever plate in time domain. Ocean Engineering, 34: 122–132.

Kara, F. (2010). Time domain prediction of power absorption from ocean waves with latching control. Renewable Energy, 35: 423–434.

Kara, F. (2011). Time domain prediction of added-resistance of ships. Journal of Ship Research, 55(3): 163–184.

Kara, F. (2015). Time domain prediction of hydroelasticity of floating bodies. Applied Ocean Research, 51: 1–13.

Kara, F. (2016). Time Domain Prediction of Seakeeping Behaviour of Catamarans. International Shipbuilding Progress, 62(3-4): 161–187.

Kara, F. (2018). Comparison of Potential and Source Methods in Time Domain with Application to Twin-hull High Speed Crafts. Ocean Engineering - Paper Accepted for Publication.

Kashiwagi, M. (2000). Hydrodynamic interactions among a great number of columns supporting a very flexible structure. Journal of Fluids and Structures, 14: 1013–1034.

Kashiwagi, M. (1993). Heave and pitch motions of a catamaran advancing in waves. Proceedings of 2nd International Conference on Fast Sea Transportations, Yokohama, Japan, 643–655.

King, B.W. (1987). Time Domain Analysis of Wave Exciting Forces on Ships and Bodies. PhD thesis, The Department of Naval Architecture and Marine Engineering, The University of Michigan. Ann Arbor, Michigan, USA.

Konispoliatis, D.N. and Mavrakos, S.A. (2013). Hydrodynamics of multiple vertical axisymmetric OWC's device srestrained in waves. *In*: 32nd International Conference on Ocean, Offshore and Arctic Engineering (OMAE 2013); Nantes, France.

Kring, D. and Sclavounos, P.D. (1991). A new method for analyzing the seakeeping of multi-hull ships. Proceedings of 1st International Conference on Fast Sea Transportation, Trondheim, Norway, 429–444.

Liapis, S. (1986). Time Domain Analysis of Ship Motions. PhD thesis, The Department of Naval Architecture and Marine Engineering, The University of Michigan, Ann Arbor, Michigan, USA.

Linton, C.M. and McIver, M. (2001). Handbook of mathematical techniques for wave-structure interactions. Chapman and Hall.

Maniar, H.D. and Newman, J.N. (1997). Wave diffraction by a long array of cylinders. Journal of Fluid Mechanics, 339: 309–330.

Mavrakos, S.A. and Mc Iver, P. (1997). Comparison of methods for computing hydrodynamic characteristics of array of wave power devices. Applied Ocean Research, 19: 283–291.

Mavrakos, S.A. (1991). Hydrodynamic coefficients for groups of interacting vertical axisymmetric bodies. Ocean Engineering, 18: 485–515.

Mavrakos, S.A., Katsaounis, G.M., Nielsen, K. and Lemonis, G. (2004). Numerical performance investigation of an array of heaving wave power converters in front of a vertical breakwater. Proceedings of 14th international offshore and polar engineering conference (ISOPE-2004), 238–245.

McIver, P. and Evans, D.V. (1984). Approximation of wave forces on cylinder arrays. Applied Ocean Engineering, 6(2): 101–107.

McIver, P. (2002). Wave interaction with arrays of structures. Applied Ocean Research, 24: 121–126.

Nader, J.-R., Zhu, S.-P., Cooper, P. and Stappenbelt, B. (2012). A finite-element study of the efficiency of arrays of oscillating water column wave energy converters. Ocean Engineering, 43: 72–81.

Nader, J.-R., Zhu, S.-P. and Cooper, P. (2014). Hydrodynamic and energetic properties of a finite array of fixed oscillating water column wave energy converters. Ocean Engineering, 88: 131–148.

Nakos, D. and Sclavounos, P.D. (1990). Ship motions by a three dimensional rankine panel method. Proceedings of the 18th Symposium on Naval Hydrodynamics. Ann. Arbor, Michigan, USA, 21–41.

Nakos, D., Kring, D. and Sclavounos, P.D. (1993). Rankine panel method for transient free surface flows. Proceedings of the 6th International Symposium on Numerical Hydrodynamics, Iowa City, I.A., USA, 613–632.

Ogilvie, T.F. (1964). Recent progress toward the understanding and prediction of ship motions. Proceedings of the 5th Symposium on Naval Hydrodynamics, Office of Naval Research, Washington, D.C., USA, 3–128.

Ohkusu, M. (1974). Hydrodynamics forces on multiple cylinders in waves. Proceedings of the International Symposium on Dynamics of Marine Vehicles and Structures in Waves, 107–112.

Ohkusu, M. (1969). On the heaving motion of two circular cylinders on the surface of a fluid. Reports of Research Institute for Applied Mechanics, No. 58, 17: 167–185.

Ohkusu, M. (1972). Wave action on groups of vertical circular cylinders. Journal of the Society of Naval Architects in Japan, 131.

Ronæss, M. (2002). Wave induced motions of two ships advancing on parallel course. PhD thesis, Department of Marine Hydrodynamics, NTNU, Trondheim, Norway.

Siddorn, P. and Eatock Taylor, R. (2008). Diffraction and independent radiation by an array of floating cylinders. Ocean Engineering, 35: 1289–1303.

Simon, M.J. (1982). Multiple scattering in arrays of axisymmetric wave-energy devices. Part 1. A matrix method using a plane-wave approximation. Journal of Fluid Mechanics, 120: 1–25.

Singh, J. and Babarit, A. (2014). A fast approach coupling Boundary Element Method and plane wave approximation for wave interaction analysis in sparse arrays of wave energy converters. Ocean Engineering, 85: 12–20.

Spring, B.H. and Monkmeyer, P.L. (1974). Interaction of plane waves with vertical cylinders. Proceedings of 14th international conference on coastal engineering, 1828–1845.

Stansby, P., Carpintero, E., Stallard, T. and Maggi, A. (2015). Three-float broad-band resonant line absorber with surge for wave energy conversion. Renewable Energy, 78: 132–140.

Taghipour, M. and Moan, T. (2008). Efficient frequency domain analysis of dynamic response for multi-body wave energy converter in multi-directional waves. Proceedings of the 18th International Offshore and Polar Engineering Conference.

Thomas, G.P. and Evans, D.V. (1981). Arrays of three-dimensional wave-energy absorbers. Journal of Fluid Mechanics, 108: 67–88.

Twersky, V. (1952). Multiple scattering of radiation by an arbitrary configuration of parallel cylinders. The Journal of the Acoustical Society of America, 24(1): 42–46.

van Oortmerssen, G. (1979). Hydrodynamic interaction between two structures floating in waves. Proceedings of the 2nd International Conference on Behaviour of Offshore Structures (BOSS'79), London, UK, 339–356.

van't Veer, A.P. and Siregar, F.R.T. (1995). The interaction effects on a catamaran travelling with forward speed in waves. Proceedings of 3rd International Conference of Fast Sea Transportation, 87–98.

WAMIT User Manual, Version 7.0, 2012.

Wei, Y., Rafiee, A., Elsaesser, B. et al. (2013). Numerical simulation of an oscillating wave surge converter. *In*: Proceedings of the 32nd international Conference on Ocean, Offshore and Arctic Engineering. Nantes, France, ASME.

Westphalen, J., Greaves, D.M., Raby, A. et al. (2014). Investigation of wave-structure interaction using state of the art CFD techniques. Open Journal of Fluid Dynamics, 4: 18.

Wolgamot, H.A. and Taylor, P.H. Taylor Eatock, R. (2012). The interaction factor and directionality in wave energy arrays. Ocean Engineering, 47: 65–73.

Xiang, X. and Faltinsen, O.M. (2011). Time domain simulation of two interacting ships advancing parallel in waves. Proceedings of the ASME 30th International Conference on Ocean, Offshore and Arctic Engineering, Rotherdam, The Netherlands.

Yilmaz, O. (1998). Hydrodynamic interactions of waves with group of truncated vertical cylinders. Journal of Waterway, Port, Coastal and Ocean Engineering, 124(5): 272–279.

Yu, Y. and Hand Li, Y. (2013). Reynolds-averaged Navier–Stokes simulation of the heave performance of a two-body floating-point absorber wave energy system. Computers & Fluids, 73: 104–114.

7

Variable Structure Control via Coupled Surfaces for Control Effort Reduction in Remotely Operated Vehicles

A. Baldini,[1] *L. Ciabattoni,*[1] *A.A. Dyda,*[2] *R. Felicetti,*[1]
F. Ferracuti,[1] *A. Freddi,*[1] *A. Monteriù*[1,*] *and D. Oskin*[3]

1. Introduction

The underwater environment is one of the most challenging for autonomous vehicles to operate in. As a consequence, in the last decades, industry and research communities showed an ever growing interest in the development of cost-effective Unmanned Underwater Vehicles (UUVs) (Yuh 2000). Among UUVs classes, Remotely Operated Vehicles (ROVs) represent a widespread and commonly used solution. Since the '50s ROVs have been employed for military purposes, especially

---

[1] Dipartimento di Ingegneria dell'Informazione, Università Politecnica delle Marche - Ancona, Italy.
[2] Admiral Nevelskoy Maritime State University - Vladivostok, Russia.
[3] Department of Information Control Systems, For Eastern Federal University & Admiral Nevelskoy Maritime State University - Vladivostok, Russia.
* Corresponding author: a.monteriu@univpm.it

by the British and American navies, due to their capabilities to recover torpedoes, mines, and even nuclear devices (Matika and Koroman 2001). In the present days, ROVs show their usefulness also in many civil operations such as monitoring submarine oil plants (Salgado 2010) and inspecting dams (Battle et al. 2003), wrecks, and submarines (Nornes et al. 2015).

In this kind of scenarios, the synthesis of a robust control system represents a key task in ROVs design, because of the nonlinearity of the model, the presence of submarine currents that act as disturbances, and the requirement of an accurate position and orientation control to operate even in narrow areas. Many works about ROV control are proposed in the literature. The most straightforward method is to design linear control laws, such as Proportional–Integral–Derivative (PID) controllers, combined with a linearized model, even though Linear Parameter-Varying (LPV) models and appropriate control techniques have been developed (Nakamura et al. 2001, Jetto and Orsini 2012). Although, model based techniques are used always more frequently due to the many studies about the mathematical model of ROVs. The most popular nonlinear techniques in this context are sliding mode (Corradini et al. 2013, Javedi-Moghaddum and Bagheri 2010, Baldeni 2017), backstepping (Fossen and Ross 2012), and dynamic surface control (Miao and Luo 2013).

The ROV is usually an overactuated system, hence there is room for control effort distribution among its thrusters. Since the '80s, the control allocation technique represents one of the most used in the field of mobile robots control (Johansen and Fossen 2013) because it greatly simplifies the synthesis of the control laws and provides to distribute the effort of the virtual inputs to the available actuators. Among the approaches which cope with the control allocation problem, the most known are: analytical and numerical minimization of a cost function under constraints (Enns 1998, Bodson 2002, Durham 1993), real-time iterative optimization (Bodson 2002, Harkegard 2002, Johansen et al. 2004), and explicit solutions for unconstrained problems (Sordalen 1997, Berge and Fossen 1997).

The main contribution of this chapter is to develop a coupled surface Variable Structure Control (VSC), following the idea proposed in (Dyda et al. 2016, Dyda et al. 2017) to solve the tracking problem for a ROV. In detail, two sliding surfaces are employed, so that the control input is required only outside the region delimited by two sliding surfaces. With respect to the classical single sliding surface control, this approach performs better in terms of energy consumption. In addition, in this chapter the problem is moved from regulation to tracking, and simulation trials prove that both energy consumption and tracking error are lower with respect to the classical sliding mode control. Fair comparison of the two controllers is ensured by adopting the same tuning strategy, based on the Artificial Bee Colony algorithm (Karaboga and Basturk 2007), a state-of-the-art heuristic algorithm that can be applied to a large class of optimization problems. Finally, thrust allocation is accounted for by linear Moore-Penrose solution (Ben-Israel and Greville 2003).

The chapter is organized as follows. The mathematical model of the ROV is introduced in Section 2. The coupled surfaces VSC technique is described in Section 3, while the control laws for the ROV in exam are discussed in Section 4. The solution to the control allocation problem is illustrated in Section 5. Simulation details and results are reported in Section 6. Conclusions and final remarks end the chapter.

2. ROV Mathematical Model

By using classical mechanics, the ROV can be modeled as a rigid body with six degrees of freedom, corresponding to the position and orientation with respect to a given coordinate system (Corradini et al. 2011, Conter et al. 1996, Longi and Rossolini 1989). Let us consider the inertial frame $R(0, x, y, z)$ and the body reference frame $R_a(0_a, x_a, y_a, z_a)$ (Conter et al. 1989), as shown in Figure 1.

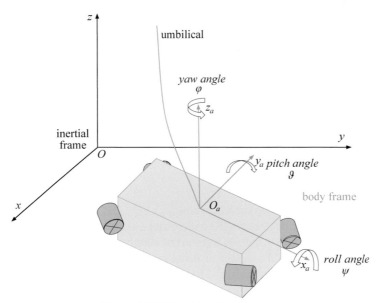

Figure 1. ROV inertial and body frames.

Using the standard notation in roll, pitch and yaw angles (i.e., ψ, θ, and φ, respectively), the ROV motion in the body-fixed frame can be expressed by the following equation (Fossen et al. 1994):

$$M\dot{v} + C(v)v + D(v)v + g(\eta) = \tau \tag{1}$$

where M is the inertia matrix (including added mass), $C(v)$ is the matrix of Coriolis and centripetal terms, $D(v)$ is the drag matrix, $g(\eta)$ is the vector of gravity, buoyancy forces and moments and τ is a vector composed by control forces and moments acting on the ROV center of mass, v is the ROV spatial velocity vector with respect to its body-fixed frame, and η is the position and orientation state vector with respect to the inertial frame. The vectors v and η can be written as follows:

$$v = [v_1 \ v_2]^T \qquad \eta = [\eta_1 \ \eta_2]^T \tag{2}$$

where: $v_1 = [u \ v \ w]^T = [\dot{x} \ \dot{y} \ \dot{z}]^T$, $v_2 = [p \ q \ r]^T = [\dot{\varphi} \ \dot{\theta} \ \dot{\psi}]^T$, $\eta_1 = [x \ y \ z]^T$ and $\eta_2 = [\varphi \ \theta \ \psi]^T$. The relationship between the velocity vector in the body-fixed frame and the position vector in the earth frame is described by:

$$\dot{\eta} = J(\eta)v = \begin{bmatrix} J_1(\eta_2) & 0_{3\times3} \\ 0_{3\times3} & J_2(\eta_2) \end{bmatrix} v$$

$$J_1(\eta_2) = \begin{bmatrix} c_\psi c_\theta & -s_\psi c_\varphi + c_\psi s_\theta s_\varphi & s_\psi s_\varphi + c_\psi c_\varphi s_\theta \\ s_\psi c_\theta & c_\psi c_\varphi + s_\varphi s_\theta s_\psi & -c_\psi s_\varphi + s_\theta s_\psi c_\varphi \\ -s_\theta & c_\theta s_\varphi & c_\theta c_\varphi \end{bmatrix} \tag{3}$$

$$J_2(\eta_2) = \begin{bmatrix} 1 & s_\varphi t_\theta & c_\varphi t_\theta \\ 0 & c_\theta & -s_\varphi \\ 0 & s_\varphi/c_\theta & c_\varphi/c_\varphi \end{bmatrix}$$

where $s_{(\cdot)} = \sin(\cdot)$, $c_{(\cdot)} = \cos(\cdot)$ and $t_{(\cdot)} = \tan(\cdot)$.

In order to take into account the underwater currents, it is possible to express the equations of the ROV motion in terms of relative velocity $v_r = v - v_c$, where $v_c = [u_c \ v_c \ w_c \ 0 \ 0 \ 0]^T$ is the vector of irrotational body-fixed underwater current velocities, hence the dynamics of the ROV motion in (1) results:

$$M\dot{v}_r + C(v_r)v_r + D(v_r)v_r + g(\eta) = \tau \tag{4}$$

Assuming that the ROV depth is independently controlled by the surface vessel, as in the considered vehicle, the ROV is able to move, thanks to its propellers, along a horizontal plane at constant depth. In particular, pitch and roll angles (i.e., θ and ψ respectively) have an irrelevant influence on the dynamical model, because their amplitude is very low, and thus they can be neglected in any operational condition (Longhi and Rossolini 1989), and the ROV model (4) can be reduced (Corradini et al. 2011). The following system of differential equations describes the ROV motion in the horizontal plane:

$$
\begin{aligned}
(M + m)\, \ddot{x} + H_x + R_x - T_x &= 0 \\
(M + m)\, \ddot{y} + H_y + R_y - T_y &= 0 \\
(I_z + i_z)\, \ddot{\varphi} + M_r + M_d + M_c - M_z &= 0
\end{aligned}
\tag{5}
$$

where M is the vehicle mass, m is the additional mass, I_z is the vehicle inertia moment around the z axis, i_z is the additional inertia moment, and M_c is the resistance moment of the cable. The quantities H_x, H_y are forces produced by the cable traction, and they can be expressed as:

$$
H_x = K(x - GV_{cx}\|V_c\|) \qquad H_y = K(y - GV_{cy}\|V_c\|)
\tag{6}
$$

where $V_c = [V_{cx}\ V_{cy}]^T$ is the velocity of the submarine current and the constants K and G can be obtained as:

$$
K = \frac{W}{\log\left(1 + \frac{WL}{T_v}\right)} \qquad G = \left(L + \frac{T_v}{K}\right)\rho_w C_{dc} \frac{D_c}{2W}
\tag{7}
$$

where L is the cable length, T_v is the vehicle weight in the water, W is the weight for length unit of the cable, ρ_w is the water density, C_{dc} is the drag coefficient of the cable and D_c is the cable diameter.

The drag forces along the x and y axes, respectively R_x and R_y in (5), are given by:

$$
\begin{aligned}
R_x &= \frac{1}{2}\rho_w V_x \|V\| (C_{d1} C_{r1} S_1 |\cos(\varphi)|) + \frac{1}{2}\rho_w V_x \|V\| (C_{d2} C_{r2} S_2 |\sin(\varphi)|) \\
R_y &= \frac{1}{2}\rho_w V_y \|V\| (C_{d1} C_{r1} S_1 |\sin(\varphi)|) + \frac{1}{2}\rho_w V_y \|V\| (C_{d2} C_{r2} S_2 |\cos(\varphi)|) .
\end{aligned}
\tag{8}
$$

In (8), C_{di} is the drag coefficient of the i-th side wall ($i = 1, 2$), C_{ri} is the packing coefficient (depending on the geometrical characteristics of the i-th side wall ($i = 1, 2$)), S_i is the area of the i-th side wall ($i = 1, 2$) and $V = [V_x\ V_y]^T = [(\dot{x} - V_{cx})\ (\dot{y} - V_{cy})]^T$.

The quantities M_d and M_r in (5) are the components of the drag torque around the z-axis produced by the vehicle rotation and by the current, respectively, and they are given by:

$$
\begin{aligned}
M_d &= \frac{1}{2}\rho_w C_d C_r S r^3\, \dot{\varphi}|\dot{\varphi}| \\
M_r &= \frac{1}{8}\rho_w \|V_c\|^2 (C_{d1}C_{r1} - C_{d2}C_{r2})\, d_1 d_2 d_3 \sin\left(\frac{\varphi - \varphi_c}{2}\right)
\end{aligned}
\tag{9}
$$

Table 1. Expressions of the model parameters. The numerical values are referred to a real ROV system, and they have been validated in (Bertozzi et al. 2001).

Parameter	Expression	Value	Unit
p_1	$M + m$	12670	kg
p_2	$\frac{1}{2}\rho_w C_{d1} C_{r1} S_1$	2667	kg/m
p_3	$\frac{1}{2}\rho_w C_{d2} C_{r2} S_2$	4934	kg/m
p_4	$W/\left[\log\left(1 + \frac{WL}{T_v}\right)\right]$	417	N/m
p_5	$(p_4 L + T_v)\rho_w C_{dc}\dfrac{D_c}{2W}$	46912	kg/m
p_6	$I_z + i_z$	18678	$kg \cdot m^2$
p_7	$\frac{1}{2}\rho_w C_d C_r S r^3$	9200	$kg \cdot m^2$
p_8	$\frac{1}{8}\rho_w [C_{d1} C_{r1} - C_{d2} C_{r2}] d_1 d_2 d_3$	−308.4	kg
p_9	M_c	1492	$N \cdot m$

where C_d is the drag coefficient of rotation, C_r is the packing coefficient of rotation, S is the equivalent area of rotation, r is the equivalent arm of action, d_i ($i = 1, 2, 3$) are the vehicle dimensions along the x_a, y_a, and z_a axes, respectively, and φ_c is the azimuth angle that describes the direction of the marine current, determined by the four quadrant inverse tangent $\varphi_c = \text{atan2}(V_{cy}, V_{cx})$. This model is in agreement with the models that are usually proposed in the literature for the motion of underwater ROVs moving in the dive plane (Fossen 1994). By substituting Equation (6)–(9) into system (5), the following equations can be obtained:

$$p_1 \ddot{x} + (p_2 \, |\cos(\varphi)| + p_3 \, |\sin(\varphi)|) \, V_x \, \|V\| + p_4 x - p_5 V_{cx} \|V_c\| = T_x$$

$$p_1 \ddot{y} + (p_2 \, |\sin(\varphi)| + p_3 \, |\cos(\varphi)|) \, V_y \, \|V\| + p_4 y - p_5 V_{cy} \|V_c\| = T_y \qquad (10)$$

$$p_6 \ddot{\varphi} + p_7 \dot{\varphi} \, |\dot{\varphi}| + p_8 \, \|V_c\|^2 \sin\left(\frac{\varphi - \varphi_c}{2}\right) + p_9 = M_z$$

The parameters p_i ($i = 1, \ldots, 9$) are known, and their nominal values are reported in Table 1. Their numerical values have been validated in (Bertozzi et al. 2001) through an analysis based on experimental data collected in real operative conditions.

2.1 State-space Model

Defining the state space variables as:

$$x_1 = x \quad x_2 = \dot{x} \quad x_3 = y \quad x_4 = \dot{y} \quad x_5 = \varphi \quad x_6 = \dot{\varphi} \qquad (11)$$

where x and y are vehicle linear positions and φ is the yaw angle, the ROV mathematical model can be rewritten in the nonlinear state space form:

$$
\begin{cases}
\dot{x}_1 = x_2 \\
\dot{x}_2 = \left[\sqrt{(V_{cx} - x_2)^2 + (V_{cy} - x_4)^2} \, (p_2 \, |\cos(x_5)| + p_3 \, |\sin(x_5)|) \, (V_{cx} - x_2) - p_4 x_1 + V_{cx} p_5 \sqrt{V_{cx}^2 + V_{cy}^2} + T_x \right] / p_1 \\
\dot{x}_3 = x_4 \\
\dot{x}_4 = \left[\sqrt{(V_{cx} - x_2)^2 + (V_{cy} - x_4)^2} \, (p_3 \, |\cos(x_5)| + p_2 \, |\sin(x_5)|) \left(V_{cy} - x_4 \right) - p_4 x_3 + V_{cy} p_5 \sqrt{V_{cx}^2 + V_{cy}^2} + T_y \right] / p_1 \\
\dot{x}_5 = x_6 \\
\dot{x}_6 = -\left[p_7 x_6 \, |x_6| - p_8 \sin \left(\frac{\varphi_c - x_5}{2} \right) (V_{cx}^2 + V_{cy}^2) + p_9 - M_z \right] / p_6
\end{cases}
\tag{12}
$$

2.2 Generalized Forces and Moments

Due to the orientation of the fixed propellers, shown in Figure 2, it follows that the forces and moments generated by the propellers on the body frame are expressed by:

$$
\begin{aligned}
T_{xa} &= (T_1 + T_2 + T_3 + T_4) \cos(\alpha) \\
T_{ya} &= (-T_1 - T_2 + T_3 + T_4) \sin(\alpha) \\
M_{za} &= (-T_1 + T_2 - T_3 + T_4) (d_x \sin(\alpha) + d_y \cos(\alpha))
\end{aligned}
\tag{13}
$$

where T_{xa}, T_{ya} are the components of the forces along the x_a-axis and y_a-axis respectively, while M_{za} is the moment along the z_a-axis; the body fixed frame is represented in Figure 1. The constants d_x and d_y are equal to a half of the ROV sides, as in Figure 2, while $\alpha = \pi/4$ is the fixed angle that describes the orientation

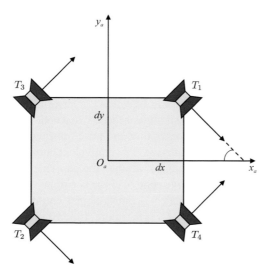

Figure 2. ROV propellers orientation.

of the propellers. The forces in the fixed frame T_x and T_y are obtained by means of a rotation of T_{xa} and T_{ya} around the z-axis:

$$T_x = cos(\varphi)T_{xa} - sin(\varphi)T_{ya}$$
$$T_y = sin(\varphi)T_{xa} + cos(\varphi)T_{ya} \qquad (14)$$
$$M_z = M_{za}$$

where T_x, T_y are the components of the forces along the x-axis and y-axis respectively, while M_z is the moment along the z-axis; the fixed inertial frame is represented in Figure 1. The complete dynamical model is obtained by substituting Equations (13) and (14) into Equation (12).

3. Coupled Surfaces VSC

Discontinuous control techniques of an appropriate class of systems, such as variable structure systems and relay systems, are characterized by a control law in the form:

$$u = \begin{cases} u^+, & s(x) > 0 \\ u^-, & s(x) < 0 \end{cases} \qquad (15)$$

where $s(x) = s(x_1, x_2, \ldots, x_n) = 0$ is a pre-designed switching surface (Figure 3(a)) in the state space with coordinates $x_1, x_2, \ldots, x_n$. The condition of sliding mode existence is (Utkin 1982):

$$s(x)\dot{s}(x) < 0 \qquad (16)$$

From a control point of view, this is equivalent to choosing $V(x) = \dfrac{1}{2}s^2(x)$ as a candidate Lyapunov function, and looking for an input u such that $\dot{V}(x) = s(x)\dot{s}(x) < 0$, and therefore the surface $s(x)$ is stable and attractive (Khalil 2001). In fact, Equation (16) implies that every system trajectory tends to the switching surface, disregarding the side of the current state with respect to the surface. Discontinuous control law in Equation (15) is fulfilled by a relay element of "sign" type. As soon as the sliding surface is reached, the relay signal starts to bounce quickly between its maximum and minimum value, and generates the well-known phenomenon of chattering. The equivalent control signal is defined in terms of average of the signal and, commonly, the equivalent control is relatively small; for this reason, an excessive control effort is employed by the sliding mode controller, while a lower (equivalent) control signal could be sufficient to keep the state inside the sliding surface. Furthermore, the fast changing polarity of the control signal is an unfavourable working condition for the actuators, maximizing the electrical and mechanical stress to the actuator itself. In general, these kinds of control signals represent a problem in real implementations, with some exceptions such as control of switching devices in power electronics.

In this chapter, the approach proposed in (Dyda et al. 2016), based on the presence of coupled sliding surfaces, is further investigated, with the aim of highlighting the differences with single sliding surface control in terms of performance indices (i.e., energy consumption and tracking errors). In particular, the control techniques are employed in a reference tracking scenario, while previous works are focused on set-point regulation.

Coupled sliding surfaces consists of two closely located switching surfaces in the system state space. This pair can be considered as any splitting of original surface $s(x) = 0$ into two surfaces $s_1(x) = 0$ and $s_2(x) = 0$, which comprises a region in the state space (or, equivalently, in the error space), as depicted in Figure 3(b). As a design choice, the control signal inside this region is assigned to zero, while outside of these sectors, the control is similar to the conventional VSC:

$$u = \begin{cases} u^+, & s_1(x) > 0,\ s_2(x) > 0 \\ 0, & s_1(x)s_2(x) < 0 \\ u^-, & s_1(x) < 0,\ s_2(x) < 0 \end{cases} \qquad (17)$$

From the phase portrait shown in Figure 3(b), it is possible to see that there is an area where the system has a free motion. Considering the free motion region in the second sector, as the error e is negative and its derivative $\dot{e}$ is positive, the state point ideally moves to the switching surface s_1, on which the sliding mode begins. The proposed switching surfaces design lets the control signal in sliding mode to be unipolar: in this case (see Figure 4), the discontinuous controller is analogous to the pulse width modulator and it results to be effective from the energetic point of view in comparison with the conventional VSC (Markin and Dyda 2000, Dyda 2003). Moreover, several switching functions can be used in order to approximate the phase trajectories of the system in case of relay control inputs: in this way, a trade-off between tracking performances and energy effectiveness can be

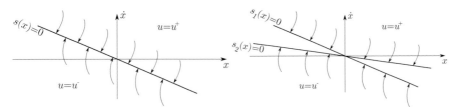

Figure 3. Sliding surfaces.

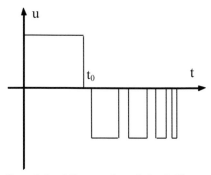

Figure 4. Control signal (in case of coupled switching surfaces).

investigated. In particular, linear sliding surfaces are proposed in this chapter, in both single surface and coupled surfaces form.

4. Controller Design

A tracking controller can be designed by considering the ROV model (12) as a composition of three interconnected Single-Input Single-Output (SISO) subsystems:

$$\begin{cases} \dot{x}_1 = x_2 \\ \dot{x}_2 = \left[\sqrt{(V_{cx} - x_2)^2 + (V_{cy} - x_4)^2}(p_2 \,|cos(x_5)| + p_3 \,|sin(x_5)|)(V_{cx} - x_2) - p_4 x_1 + V_{cx} p_5 \sqrt{V_{cx}^2 + V_{cy}^2} + T_x \right] / p_1 \end{cases}$$

$$\begin{cases} \dot{x}_3 = x_4 \\ \dot{x}_4 = \left[\sqrt{(V_{cx} - x_2)^2 + (V_{cy} - x_4)^2}(p_3 \,|cos(x_5)| + p_2 \,|sin(x_5)|)(V_{cy} - x_4) - p_4 x_3 + V_{cy} p_5 \sqrt{V_{cx}^2 + V_{cy}^2} + T_y \right] / p_1 \end{cases}$$

$$\begin{cases} \dot{x}_5 = x_6 \\ \dot{x}_6 = -\left[p_7 x_6 \,|x_6| - p_8 \sin(\frac{\varphi_c - x_5}{2})(V_{cx}^2 + V_{cy}^2) + p_9 - M_z \right] / p_6 \end{cases} \tag{18}$$

The state variables of the first subsystem are x_1 and x_2, which represent, respectively, the position x and the velocity $\dot{x}$ along the same axis, while the control input T_x is the generalized force along the x-axis in the earth fixed-frame. In other words, the first subsystem represents the ROV dynamics along the x-axis. Similarly, the second subsystem represents the ROV dynamics along the y-axis. In fact, the state variables are x_3 and x_4, namely the position y and the velocity $\dot{y}$ along the y-axis, while the control input T_y is the generalized force along the y-axis in the earth fixed-frame. Finally, the yaw dynamics is captured by the third subsystem, where the state variables are x_5 (φ) and x_6 ($\dot{\varphi}$) and the control input M_z is the generalized torque along the z-axis in the earth fixed-frame. In the following, the three subsystems will be referred to as subsystem Σ_x, Σ_y, and Σ_φ respectively.

It should be remarked that the considered subsystems are interconnected, i.e., all the state variables $x_1, \ldots, x_6$ can appear in every subsystem. This is not a problem under the assumption that each state variables is measured (i.e., the state is accessible), so that the influence of the non-state variables of each subsystem can be rejected as a measurable matching disturbance. From a practical point of view, this means that the equipped sensors should provide the linear/angular speeds and accelerations: such requirement is satisfied with the use of standard equipment such as an Attitude and Heading Reference System.

Another remark should be pointed out. The decomposition of the system into three different subsystems permits to design simpler controllers, which are three SISO controllers in this case. Such decomposition often leads to a better understanding of the system which can be a crucial point for engineering applications. This decomposition, however, is not mathematically correct in general. Anyway, for the ROV system (12) with the aforementioned assumptions, the subdivision leads to an exact MIMO controller. Indeed, if for each subsystem $i = x$, y, φ there exists a control law such that the Lyapunov function has definite negative derivative ($\dot{V}_i < 0$), then by choosing the candidate Lyapunov function $V = \sum V_i$, it is straightforward to show that $\dot{V} < 0$ (i.e., the whole system is asymptotically stable).

Following the theory of the discontinuous control system with coupled switching surfaces introduced in Section 3, a tracking controller can be designed by introducing two linear sliding surfaces for each subsystem, for a total of six linear sliding surfaces:

$$\begin{cases} s_{1i} = \dot{e}_i + k_{1,i} e_i \\ s_{2i} = \dot{e}_i + k_{2,i} e_i \end{cases} \tag{19}$$

where the index $i - x, y, \varphi$ specifies the subsystem Σ_i, and the errors e_x, e_y, and e_φ are defined as follows:

$$\begin{aligned} e_x &= x_d - x \\ e_y &= y_d - y \\ e_\varphi &= \varphi_d - \varphi \end{aligned} \tag{20}$$

The quantities x_d, y_d, and φ_d are the desired trajectories, while the control parameters $k_{1,i}$, $k_{2,i}$ are related by the following relation:

$$k_{1i} = \gamma_i k_{2i} \tag{21}$$

where k_{2i}, $i = x, y, \varphi$, are project parameters that define the slope of $s_2(x) = 0$ (see Figure 3(b)); $\gamma_i > 1$, $i = x, y, \varphi$, are project parameters as well, and indicate the ratio between the slopes of the two sliding surfaces $s_1(x) = 0$ and $s_2(x) = 0$. The parameters k_{2i} and $\gamma_i > 1$, $i = x, y, \varphi$ must be tuned properly in order to obtain the desired performances, hence, in this work, they are given by the Artificial Bee Colony algorithm, as illustrated in Section 6.

The following single sliding surfaces are reported and will be used in the simulation for the aim of comparison:

$$s_i = \dot{e}_i + k_i e_i \qquad (22)$$

with $i = x, y, \varphi$, where k_i are project parameters, that are also tuned by the Artificial Bee Colony algorithm.

The control vector has the following form:

$$\boldsymbol{u} = [u_x \ u_y \ u_\varphi]^T = [T_x \ T_y \ M_z]^T \qquad (23)$$

where

$$u_i = \frac{u_{0i}}{2} \left[\text{sign}(s_{1i}) + \text{sign}(s_{2i}) \right] \qquad (24)$$

for $i = x, y, \varphi$. The amplitudes of the control signals u_{0x}, u_{0y}, and $u_{0\varphi}$ are constants, whose values are chosen such that saturation of the actuators is avoided.

5. Control Allocation

Control laws are designed in term of the generalized desired forces $T_{x,d}$, $T_{y,d}$, and moment $M_{z,d}$, but they cannot be applied directly from the thrusters due to their angle, so a crucial part of the control scheme consists in calculating the desired force for each thruster. Moreover, once $T_{x,d}$, $T_{y,d}$, $M_{z,d}$ are determined by the control laws, there can be multiple ways to obtain them through the four actuators. In other words, the problem is the distribution of the control effort (in terms of generalized forces and moments) among the available thrusters.

A thruster is essentially composed of a motor and a propeller, thus it can be approximated by a linear system that relates the input (e.g., voltage, current, fuel, etc.) and the output, which is the produced force (i.e., T_i in the model). A scheme for a generic thruster is presented in Figure 5.

Considering the case of a feedback-controlled thruster, which is composed of a physical thruster and a low level controller, the proposed method does not entail any modification of this controller, hence it can be applied even to thrusters with an

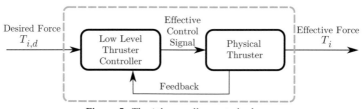

Figure 5. The i-th propeller control scheme.

embedded low level controller. According to this policy, the only available degree of freedom is the choice of the control references $T_{i,d}$.

The relation between the actual generalized forces and moments T_x, T_y, M_z and the actual forces generated by the propellers T_1, T_2, T_3, T_4, which is based on the relations (13) and (14), can be thus rewritten in the form:

$$T_x = \cos(\varphi - \alpha)(T_1 + T_2) + \cos(\varphi + \alpha)(T_3 + T_4)$$

$$T_y = \sin(\varphi - \alpha)(T_1 + T_2) + \sin(\varphi + \alpha)(T_3 + T_4) \tag{25}$$

$$M_z = (d_x \sin(\alpha) + d_y \cos(\alpha))(-T_1 + T_2 - T_3 + T_4)$$

This relation can be expressed as a matrix product by defining:

$$u = [T_1 \ T_2 \ T_3 \ T_4]^T \tag{26}$$

$$\tau = [T_x \ T_y \ M_z]^T \tag{27}$$

$$d_a = (d_x s_a + d_y c_a) \tag{28}$$

where $\alpha = \pi/4$, $c_{(\varphi-\alpha)} = \cos(\varphi - \alpha)$, $c_{(\varphi+\alpha)} = \cos(\varphi + \alpha)$, $s_{(\varphi-\alpha)} = \sin(\varphi - \alpha)$ and $s_{(\varphi+\alpha)} = \sin(\varphi + \alpha)$. Hence, Equation (25) can be rewritten as:

$$\tau = B(\varphi) \cdot u = \begin{bmatrix} c_{(\varphi-\alpha)} & c_{(\varphi-\alpha)} & c_{(\varphi+\alpha)} & c_{(\varphi+\alpha)} \\ s_{(\varphi-\alpha)} & s_{(\varphi-\alpha)} & s_{(\varphi+\alpha)} & s_{(\varphi+\alpha)} \\ -d_a & d_a & -d_a & d_a \end{bmatrix} \cdot u \tag{29}$$

Similarly, u_d and τ_d can be defined as follows:

$$u_d = [T_{1,d} \ T_{2,d} \ T_{3,d} \ T_{4,d}]^T \tag{30}$$

$$\tau_d = [T_{x,d} \ T_{y,d} \ M_{z,d}]^T \tag{31}$$

The relationship between virtual forces and actual thrusts can be described by linear equations; as the adopted control technique is a relay control, there is no reason to consider constraints such as saturations because they are intrinsically avoided through a proper choice of the amplitude of the control signal. As a consequence, thrusters' transients and saturations are neglected (hence, $u = u_d$), and the problem is formulated as follows. Given a set of desired control outputs τ_d generated by the control laws, find a set of references u_d, such that:

$$\tau_d = \tau = B(\varphi) \cdot u = B(\varphi) \cdot u_d \tag{32}$$

The first equation is consequent to the aim of the allocation: the objective is to obtain that the actual generalized forces and moment are equal to the desired ones.

The second equation is directly obtained substituting Equation (29), while the third equation follows from the previous assumption. Ideally, the control allocation should map the two aforementioned domains, but this relation cannot be univocal, since the domains have different dimensions, i.e., the desired mapping is from a three dimensional space into a four dimensional space. In particular, since the relation between T_x, T_y, M_z and T_1, T_2, T_3, T_4 is linear, multiple solutions could exist, except for some singular geometrical configurations of the ROV, that reduce the rank of the thruster configuration matrix, as shown in (Ciabattoni et al. 2016). It follows that there is a thruster redundancy, and the system is over-actuated, so a minimization of a cost function is possible. In order to keep a low complexity of the control allocation algorithm, a common choice is to minimize the cost function $J = \sum_{i=1}^{4} T_i^2 = \sum_{i=1}^{4} T_{i,d}^2$, which roughly represents the energy consumption. Moreover, since the energy consumption is related to the forces produced by the actuators, it leads to a moderate control effort, with the result of lowering the risk of saturation of the propellers. However, saturation is not a concern in this control scheme, as the amplitudes of the generalized forces and moments u_{0x}, u_{0y} and $u_{0\varphi}$ are chosen such that they are always feasible (in absence of faults and failures). The problem can be solved with a Moore-Penrose based solution (Ciabattoni et al. 2016), which returns the least-square solution for linear unconstrained problems. Although $B(\varphi)$ is time varying, the control allocation problem can be instantly considered as a linear system of equations for each sample time, due to the fact that φ is measured. The solution is thus given by:

$$u_d = B^\dagger(\varphi) \cdot \tau_d \tag{33}$$

where $B^\dagger(\varphi)$ is the Moore-Penrose pseudoinverse of $B(\varphi)$. In particular:

- if multiple solution for Equation (29) exists, Equation (33) solves the problem:

$$\min_{u_d} \|u_d\| \quad \text{s.t.} \quad B(\varphi) \cdot u_d = \tau_d \tag{34}$$

  where $B^\dagger(\varphi) = (B^T(\varphi)B(\varphi))^{-1} B^T(\varphi)$.
- if only one solution exists (i.e., the matrix is invertible), Equation (33) corresponds to $u_d = B^{-1}(\varphi) \cdot \tau_d$;
- if no solution of (29) exists, Equation (33) solves the mean square error problem:

$$\min_{u_d} \|B(\varphi) \cdot u_d - \tau_d\| \tag{35}$$

  where $B^\dagger(\varphi) = B^T(\varphi) (B(\varphi)B^T(\varphi))^{-1}$.

The overall proposed scheme is reported in Figure 6.

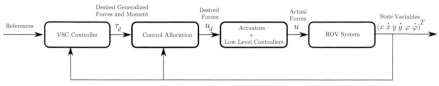

Figure 6. Control scheme.

6. Simulation Results

In order to verify and test the proposed control method, several simulations have been performed where the main aspects are summarized as follows:

- each actuator is composed by an electric motor and a propeller where the latter is fixed on the rotational axis of the related motor. In some applications, where the actuator dynamics is resolutely lower than the system dynamics, the transient of the control variable can be neglected. However, in the considered case, the high water resistance acting on the propeller is transferred to the motors which can cause a loss of performance. Hence, each actuator is modelled as a first order linear system which introduces a phase delay in the feedback. Moreover, the maximum angular speed of each motor is fixed, and thus the maximum thrust is limited. In the considered system, the maximum deliverable thrust of each propeller is $5000N$. As a consequence, the constants u_{0x}, u_{0y} and $u_{0\varphi}$ in Equation (24) are defined as $u_{0x} = u_{0y} = 4000$ and $u_{0\varphi} = 10000$, with the aim of avoiding thruster saturation;

- the fundamental sampling time for the simulation is fixed to 5 *ms*, while the controller output is calculated every 10 *ms*.

A random heuristic search algorithm is employed in order to tune the control laws parameters, with the aim of comparing in a fair way the performances of the single surface and the coupled surfaces variable structure controllers. The Artificial Bee Colony algorithm (Karaboga and Basturk 2007) is a proper choice because it allows to numerically minimize any type of cost function and can be applied to a wide class of problems. The physical relation between the power consumption and the thrust is expressed by (Johansen et al. 2004):

$$P(t) = k_1|T_1(t)|^{3/2} + k_2|T_2(t)|^{3/2} + k_3|T_3(t)|^{3/2} + k_4|T_4(t)|^{3/2} \tag{36}$$

where k_1, k_2, k_3, k_4 are constants that bind the power to the thrust for each propeller. Since there are no differences between the propellers, it is possible to assume that $k_1 = k_2 = k_3 = k_4 = \bar{k}$, for an arbitrary positive constant $\bar{k}$. Hence, the cost function can be chosen as:

$$C = \begin{cases} \int_0^T \frac{P(t)}{k} \, dt & (e_{x,max}, e_{y,max}, e_{\varphi,max}) \in [-0.2, 0.2] \times [-0.2, 0.2] \times [-2, 2] \\ +\infty & otherwise \end{cases} \qquad (37)$$

where T is the simulation total time and:

$$e_{x,max} = \max_t |x(t) - x_d(t)|$$

$$e_{y,max} = \max_t |y(t) - y_d(t)| \qquad (38)$$

$$e_{\varphi,max} = \max_t \left(\frac{180}{\pi} |\varphi(t) - \varphi_d(t)| \right).$$

In fact, the objective is the investigation of the energy consumption of the proposed techniques and, at the same time, comply with an acceptable tracking error, whose specifications are detailed in (Baldini et al. 1999).

In Figure 7, the tracking performances of the single surface (grey line) and the coupled surfaces (black line) sliding mode controllers are reported. The tracking reference (grey dashed line) is a regular hexagon of 6 *m* side, which imposes an anticlockwise rotation. The initial position of the ROV and the starting point of the trajectory coincide and such point is depicted with a grey triangle, while the orientation reference is defined such that the system is aligned with the trajectory (i.e., the ROV moves forward). It is possible to note that both controllers follow the

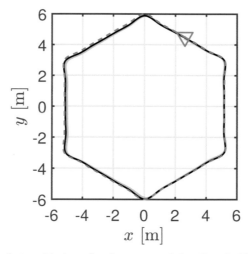

Figure 7. Tracking trajectory (single surface in grey, coupled surfaces in black, and reference in dashed grey).

reference, hence the positioning specifications are respected, as shown in Figure 8. The tracking errors along *x*-axis and *y*-axis are both acceptable. Moreover, due to the hexagonal reference symmetry, it can be seen that the errors $x_d - x$ and $y_d - y$ are quantitatively and qualitatively comparable. Single sliding surface technique involves a greater amplitude of the positioning error. Analogously, as regards the orientation error, both techniques satisfy the specifications. In particular, the single sliding surface error is strictly greater than the coupled surfaces' one. For every component, the largest errors occur when the related reference varies more widely.

The generalized forces and moment are reported in Figure 9, where the output of the discontinuous control laws is shown. It is possible to note that, as regard the single sliding surface, the control signals T_x and T_y are always bipolar, hence, the jump is between ±4000 *N*. However, the coupled sliding surfaces generates both bipolar and unipolar jumps. The main advantage is that in some time intervals the required effort is null, with consequent energy saving. Moreover, when the control signal is unipolar, the jump width is reduced by half. Similar considerations can be made about the moment M_z, although this is less evident from the same figure.

The actual control efforts are reported in Figure 10. It should be noted that the saturation limits are never reached. Figure 11(a) shows the IAE index defined in terms of:

$$IAE = \int_0^T (|x - x_d| + |y - y_d|)dt + \frac{1}{10} \frac{180}{\pi} \int_0^T |\varphi - \varphi_d| dt \tag{39}$$

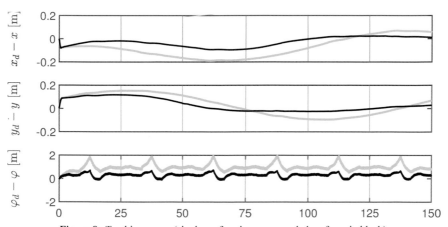

Figure 8. Tracking errors (single surface in grey, coupled surfaces in black).

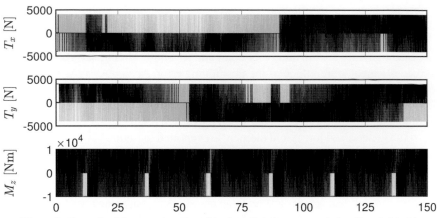

Figure 9. Generalized forces and moment (single surface in grey, coupled surfaces in black).

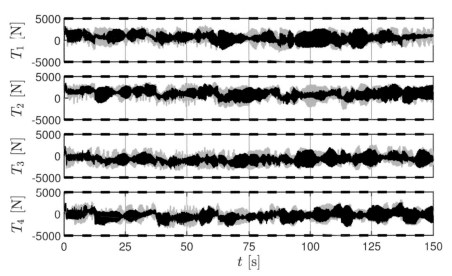

Figure 10. Control efforts (single surface in grey, coupled surfaces in black).

where the coefficient 1/10 aims to weigh equally an error of 2° of the yaw angle and an error of 0.2 m along x or y axis. The cumulative energy consumption, defined as the integral over time of (36), is reported in Figure 11(b). As expected, the consumption in presence of coupled surfaces is lower due to the unipolarity of the control signals.

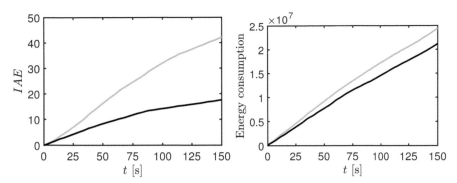

Figure 11. Performance indices (single surface in grey, coupled surfaces in black).

7. Conclusion and Outlook

In this chapter, a Variable Structure Control via coupled surfaces for a ROV has been proposed and tested in simulation. The idea behind the technique is to leave the system in free motion when the tracking error is acceptable. Due to intrinsic slowness of the system, it is reasonable to expect energy savings without significant loss of performances. From an implementation point of view, the proposed technique follows the philosophy of the SMC, with the difference that there are two sliding surfaces. Each sliding surface is linear and designed in terms of tracking error, hence, the system is in free motion when the state vector is between them. Moreover, for a fair comparison with classical SMC, an extensive tuning phase with the ABC algorithm is considered. Simulation results show that, not only energy savings is achieved, but also an improvement of tracking performance is obtained, in terms of IAE. In particular, the energy consumption is estimated with a relative cost function, which is widespread in literature. The improvement can be attributed to two main causes. Firstly, the VSC signals are null in some time intervals, with no energy consumption in there. Secondly, there is a reduction of the jump caused by the switching side of the surface. In fact, the classical SMC signals are bipolar, and thus they bounce from u^+ to u^- and vice versa. On the other hand, the control signals of the proposed VSC jump from 0 to u^+ (or u^-), and vice versa, showing a unipolar trend with a lower control effort.

Some future works are possible. First of all, it is possible to investigate about the performances of the proposed VSC considering a non bang-bang type of controller. Furthermore, it is possible to extend the control in the fault and failure scenario developing a fault tolerant control system.

References

Baldini, A., Ciabattoni, L., Fasano, A., Felicetti, R., Ferracuti, F., Freddi, A. and Monteriù, A. (2017). Active fault tolerant control of remotely operated vehicles via control effort redistribution. pp. 1–7. *In*: Proceedings of 13th ASME/IEEE International Conference on Mechatronic & Embedded Systems & Applications (MESA). Cleveland, Ohio, USA.

Baldini, M., Corradini, M., Jetto, L. and Longhi, S. (1999). A multiple-model based approach for the intelligent control of underwater remotely operated vehicles. IFAC Proceedings Volumes, 32(2): 8291–8296.

Batlle, J., Nicosevici, T., Garcia, R. and Carreras, M. (2003). ROV-aided dam inspection: Practical results, 36: 271–274.

Ben-Israel, A. and Greville, T.N. (2003). Generalized Inverses: Theory and Applications, Vol. 15, Springer Science & Business Media.

Berge, S.P. and Fossen, T.I. (1997). Robust control allocation of overactuated ships; experiments with a model ship. pp. 166–171. *In*: Proc. 4th IFAC Conf. Manoeuvring and Control of Marine Craft, Citeseer.

Bertozzi, C., Ippoliti, G., Longhi, S., Radicioni, A. and Rossolini, A. (2001). Multiple models control of a remotely operated vehicle: Analysis of models structure and complexity. IFAC Proceedings Volumes, 34(7): 437–444.

Bodson, M. (2002). Evaluation of optimization methods for control allocation. Journal of Guidance, Control, and Dynamics, 25(4): 703–711.

Ciabattoni, L., Fasano, A., Ferracuti, F., Freddi, A., Longhi, S. and Monteriù, A. (2016). A thruster failure tolerant control scheme for underwater vehicles. pp. 1–6. *In*: 12th IEEE/ASME International Conference on Mechatronic and Embedded Systems and Applications (MESA).

Conter, A., Longhi, S. and Tirabassi, C. (1989). Dynamic model and self-tuning adaptive control of an underwater vehicle. 8th Int. Conf. on Offshore Mechanics and Arctic Engineering, Hague, Netherlands, pp. 139–146.

Corradini, M., Monteriù, A., Orlando, G. and Pettinari, S. (2011). An actuator failure tolerant robust control approach for an underwater remotely operated vehicle. pp. 3934–3939. *In*: 50th IEEE Conf. on Decision and Control and European Control Conf. Orlando, Florida, USA.

Corradini, M.L., Monteriù, A. and Orlando, G. (2011). An actuator failure tolerant control scheme for an underwater remotely operated vehicle. IEEE Transactions on Control Systems Technology, 19(5): 1036–1046.

Durham, W.C. (1993). Constrained control allocation. Journal of Guidance, Control, and Dynamics, 16(4): 717–725.

Dyda, A., Oskin, D., Longhi, S. and Monteriù, A. (2016). A nonlinear system with coupled switching surfaces for remotely operated vehicle control. IFAC-PapersOnLine, 49(23): 311–316.

Dyda, A., Oskin, D., Longhi, S. and Monteriù, A. (2017). An adaptive VSS control for remotely operated vehicles. International Journal of Adaptive Control and Signal Processing, 31(4): 507–521.

Dyda, A.A. (2003). Adaptive variable-structure system. Patent of Russian Federation No. 2210170 (In Russian). Inventions bulletin No. 22.

Enns, D. (1998). Control allocation approaches. pp. 98–108. *In*: Proceedings of AIAA Guidance, Navigation, and Control Conference.

Fossen, T. and Ross, A. (2006). Nonlinear modelling, identification and control of UUVs. IEE Control Engineering Series, 69: 13–42.

Fossen, T.I. (1994). Guidance and Control of Ocean Vehicles. J. Wiley & Sons, New York.

Harkegard, O. (2002). Efficient active set algorithms for solving constrained least squares problems in aircraft control allocation. *In*: Proceedings of the 41st IEEE Conference on Decision and Control, 2: 1295–1300.

Javadi-Moghaddam, J. and Bagheri, A. (2010). An adaptive neuro-fuzzy sliding mode based genetic algorithm control system for under water remotely operated vehicle. Expert Systems with Applications, 37(1): 647–660.

Jetto, L. and Orsini, V. (2012). A supervised switching control policy for LPV systems with inaccurate parameter knowledge. IEEE Transactions on Automatic Control, 57(6): 1527–1532.

Johansen, T.A. and Fossen, T.I. (2013). Control allocation—a survey. Automatica, 49(5): 1087–1103.

Johansen, T.A., Fossen, T.I. and Berge, S.P. (2004). Constrained nonlinear control allocation with singularity avoidance using sequential quadratic programming. IEEE Transactions on Control Systems Technology, 12(1): 211–216.

Karaboga, D. and Basturk, B. (2007). A powerful and efficient algorithm for numerical function optimization: artificial bee colony (ABC) algorithm. Journal of Global Optimization, 39(3): 459–471.

Khalil, H. (2001). Nonlinear Systems [3rd Edition], Pearson.

Longhi, S. and Rossolini, A. (1989). Adaptive control for an underwater vehicle: Simulation studies and implementation details. Proceedings of the IFAC Workshop on Expert Systems and Signal Processing in Marine Automation, pp. 271–280.

Markin, V. and Dyda, A.A. (2000). Two-switching surfaces adaptive control in variable-structure systems. pp. 48–50. In: Pacific Science Review, Vol. 2, Vladivostok, Dalnauka.

Matika, D. and Koroman, V. (2001). Undersea Detection of Sea Mines, Tech. rep. Ministry of Defence Zagreb (Croatia).

Miao, B., Li, T. and Luo, W. (2013). A DSC and MLP based robust adaptive NN tracking control for underwater vehicle. Neurocomputing, 111: 184–189.

Nakamura, M., Kajiwara, H. and Koterayama, W. (2001). Development of an ROV operated both as towed and self-propulsive vehicle. Ocean Engineering, 28(1): 1–43.

Nornes, S.M., Ludvigsen, M., Ødegard, Ø. and SØrensen, A.J. (2015). Underwater photogrammetric mapping of an intact standing steel wreck with ROV. IFAC-PapersOnLine, 48(2): 206–211.

Salgado-Jimenez, T., Gonzalez-Lopez, J., Martinez-Soto, L., Olguin-Lopez, E., Resendiz-Gonzalez, P. and Bandala-Sanchez, M. (2010). Deep water ROV design for the Mexican oil industry. pp. 1–6. In: Proceedings of the IEEE OCEANS.

Sørdalen, O. (1997). Optimal thrust allocation for marine vessels. Control Engineering Practice, 5(9): 1223–1231.

Utkin, V.I. (1981). Principles of identification using sliding regimes. Soviet Physics Doklady, 26: 271–272.

Yuh, J. (2000). Design and control of autonomous underwater robots: A survey. Autonomous Robots, 8(1): 7–24.

8

Marine Brushless A.C. Generators

Zenghua Sun[1] and *Guichen Zhang*[2]

1. Introduction

Marine brushless A.C. generators have been designed so as to fulfil various laws and regulations. They cause the output current of a rotary armature type A.C. exciter, mounted on the shaft of the main generator, to excite the field system of the main generator through a rotary rectifier, thus generating A.C. power. Housing configuration and protecting system of the generators are as follows: drip-proof, protected, with sleeve bearings in end brackets, forced lubrication, with a built-in exciter (Zhang 2013). The static exciting systems are composed of a current transformer, a rectifier, a surge protector, a power transformer, etc. Normally, their main components are mounted as a unit on top of the generators. TAIYO model's FEK generator is showed in Figure 1.

2. Generators Construction

As shown in the structural drawing (Figure 1), the generator is composed mainly of a stator frame, stator core, stator winding, field core, field winding, shaft, bearings, A.C. exciter, rotary rectifier, and so forth. Cool air is taken in by the fan fitted to the rotor on the driven side through the ventilation port on the non-coupling side,

[1] Marine Engineering College, Dalian Maritime University, No. 1 Linghai Road, Dalian, Liaoning, P. R. China.
[2] Merchant Marine College, Shanghai Maritime University, No. 1550 Haigang Avenue, Pudong New Area, Shanghai, P. R. China.

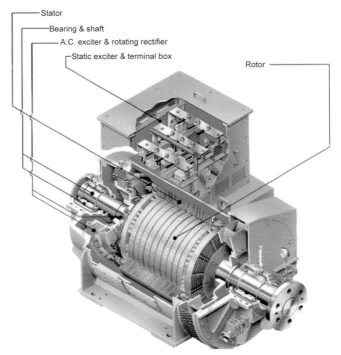

Figure 1. Marine brushless A.C. generator.

and passes over the surfaces of the field core and coil ends and through the air ducts arranged in the stator core and field core, thus effectively absorbing the heat generated by them. The heated air is let out of the generator through the ventilation portion on the coupling side. The A.C. exciter overhangs or is incorporated in the generator on the non-coupling side. The rotary rectifier is also mounted on the same side to supply an exciting current to the field winding of the generator (Jiang 2005).

2.1 Stators

The stator frames are constructed of welded mild steel plates, being so designed as to have sufficient mechanical strength and to withstand electric shock. The stator core is provided as follows: silicon steel plate which has magnetic characteristics and is coated with insulating varnish for prevention of eddy current is punched, and the punched plate elements are piled along the inner circumference of the stator frame from one side and equipped with air duct at each of regular pile intervals. The core, thus formed, is forced in and fastened by stator clampers made of steel plate. The stator winding is formed of electric wire of insulation class "F", and placed in the slots which have been formed in the inner periphery of the stator core and protected with an insulating material of class "F". The winding, thus placed, is fastened to the stator core by special wedges and then subjected to sufficient varnish impregnation and drying (Xue 2011). The outward appearance of the stator is shown in Figure 2.

Figure 2. Stator.

2.2 Rotors

The field core is made up of laminations of a material having an exceedingly high coercive force for ease of voltage self-establishment. The laminated core is fitted into the shaft (or the spider) and clamped at both ends by means of rotor clampers, and also serves to protect the winding. The field core has such cross-sectional configuration as illustrated in Figure 3. The slots in which the field coils are to be placed are formed in a number of slot groups which correspond to the number of poles. The pole center exists between slot groups as shown in Figure 3. The field winding is formed of electric wire of insulation class "F" and placed in the slots furnished with a class "F" insulating material (MV COSCO XIAMEN 2004). The

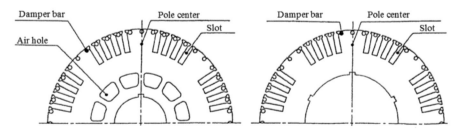

Figure 3. Core for 8 poles illustrated.

186

coils placed in the slots are fixed firmly to the field core by special wedges and subjected to sufficient varnish impregnation and drying.

The coil ends are bound by piano wire or special tape so as to not move out by centrifugal force. The external appearance of the rotor is shown in Figure 4. The shafts is made of an excellent forged steel material in careful consideration of mechanical strength. For the generators to be directly coupled to a diesel engine, special attention has been paid to the prevention of shaft breakage due to torsional vibration (MV COSCO BELGIUM 2013a).

Figure 4. External Appearance of the Rotor.

2.3 Bearings

The bearings are sleeve bearings. The sleeve bearings are split into two for convenience of disassembly and reassembly. Supported by the lower portions of end brackets and cases, the bearings have sufficient strength to withstand external forces, axial load, and vibration. Keep the bearing oil supply pressure as follows: 0.15~0.25 Mpa: 8,10P Generator; 0.10~0.20 MPa: 4,6P generator.

2.4 Ventilation

In the generators of the enclosed and self-ventilated type, the fan fitted to the rotors on the non-driven side causes cooling air to be drawn in through the ventilation port on the non-coupling side and to pass over the surfaces of the stator core and the coil ends and through the air ducts provided within the stator core and the field core. Thus, the air effectively receives heat generators by these components and then flows out through the vent port on the coupling side. The ventilation ports are provided to shield the human body and guard it from foreign solid objects.

2.5 A.C. Exciters

The A.C. exciters of the rotating armature type and located on the non-coupling side of the generators. The exciters are composed mainly of a stator frame, field core, field windings, armature core, armature winding, etc.

2.5.1 Stators

The stator frames are constructed of cast iron and welded mild steel plates, being so designed so as to have satisfactory rigidity and strength. The field core is made by placing laminations of a material having excellent magnetic characteristics in the stator frames (MV COSCO BELGIUM 2013b). In the field core, slots in which the field winding is to be placed are formed in a number of slot groups which correspond to the number of poles. The field winding is formed of electric wire of insulation class F and placed in the slots formed along the periphery of the field core and furnished with an insulating material of the employed class. The winding put in the slots is subjected to sufficient varnish impregnating and drying and fastened to the field core by wedges.

2.5.2 Rotors

A.C. exciter Rotor is as shown in Figure 5. The armature core (rotor core) is provided as follows. Silicon steel plate of high quality which is coated with an insulating varnish for prevention of eddy current is punched, and the punched plate elements are piled on the shaft and fastened by end plate. The armature winding is

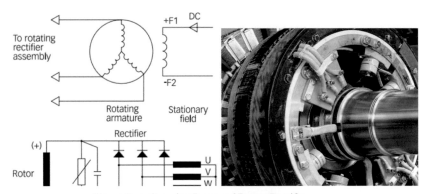

Figure 5. A.C. exciter Rotor and Rotary Rectifier.

formed of electric wire of insulation class F and placed in the slots furnished with an insulating material of the employed class (MV COSCO BELGIUM 2013b). The winding thus put in the slots is fixed to the armature core by wedges and subjected to sufficient varnish impregnation and drying. The coil ends are bound by piano wire or a special sort of tape so as not to move out by centrifugal force.

2.6 Rotary Rectifiers

The rotary rectifier is as shown in Figure 5; it is fitted to the generator shaft. Two conductive plates are attached to a rotor cramped through an insulating plate. The conductive plates carry silicon rectifiers for each phase, a silicon carbide varistor (Silistor) for surge prevention and other parts (Huang and Wang 2006). These components serve for three-phase full wave rectification of the A.C. output of the A.C. exciter to supply a D.C. exciting current to the field winding of the generator.

2.7 Space Heaters

The generator windings may absorb moisture while the generator is at rest. In order to prevent this moisture absorption, it is necessary to keep the temperature within the resting generator a little higher than the ambient temperature. A space heater is provided to satisfy this requirement. The space heater is located at the lower portion of the generator frame so that the heated air circulates in the generator to serve effectively to protect the windings and the related parts from moisture. Therefore, when it is intended to keep the generator at rest, be sure to turn on the power switch for the space heater. When starting the generator, be sure to cut off the power supply to the space heater.

2.8 Air Filters

When drip-proof, protected type generators are equipped with an-air filter, it is located at the cooling air suction port so as to minimize the entry of dust, oil vapor and other foreign matter into the generators, so that the generators interior can be kept cleaner than otherwise.

3. Exciting Systems

The basic circuit composition of this brushless generator is shown in Figure 6. The output current of the revolving-armature type A.C. exciter installed on the shaft of the rotating machine of the main generator excites the field system of the main

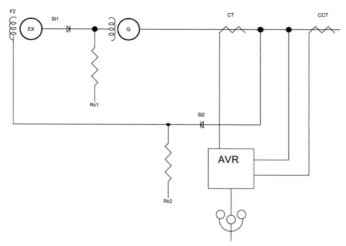

Figure 6. (a) Basic Circuit Diagram.

generator through the rotary rectifier on the shaft (MV COSCO BELGIUM 2013b). Also, the exciting current of the A.C. exciter is taken out as the exciting current for obtaining compound characteristics from part of the output of the main generator through the static exciter or current transformer (CT) and there actor (RT). This is then transformed into DC by the silicon rectifier (Si) to excite the field system of the A.C. exciter. The output of the CT and the RT is then preset at overcompensation for maintaining the terminal voltage of the generator at a constant level (Zhang and Wu 2013). This is controlled by the automatic voltage regulator (AVR) which constitutes the diversion system.

Where, G-Main generator; Ex-A.C. exciter; Si_2-Silicon rectifier; CT-Current transformer; F_2-Exfield winding; $Rc_{1,2}$-Discharge resistor; CCT-Current transformer for cross-current compensation; RT-Reactor; AVR-Automatic voltage regulator; VR-Rheostat for voltage setting; Si_1-Rotary rectifier; F_1-Main generator field system.

The excitation circuit of this generator is as shown in Figure 6(b). This system converts the output current of the A.C. exciter installed on the shaft of the main generator into DC. This is performed by the rotary rectifier and consequently excites the field system of the main generator (Zhang and Ma 2011). Exciting the A.C. rectifier is performed by taking out part of the output of the main generator as excitation current by the reactor (RT) and the current transformer (CT). This is to say that the excitation current compensates for the terminal voltage at the no-load condition. This excitation current is obtained from the RT and compensates for voltage drop by leak reactance with the armature reaction caused by load current at load being obtained from CT.

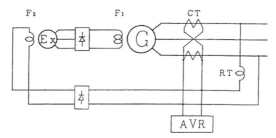

Figure 6. (b) The basic circuit diagram of the brushless A.C. generator.

4. Automatic Voltage Regulators (AVR)

4.1 Composition of AVR

AVRs is integrated into a panel as shown in Figure 7(a)~(d). The AVR is consisted of detection transformer, voltage detection circuit, micro-processor (MPU), power

Figure 7. (a) AVR (SPRESY 15); (b) AVR (TAIYO).

Figure 7. (c) AVR (for generator top mounting); (b) AVR (for panel mounting).

circuit, display circuit, pulse amplification circuit, magnetic field current control circuit, A/D for voltage setting circuit, photo-couplers and the like (Zhang and Ma 2011). Figure 8 is a block diagram showing circuit composition and signal flow.

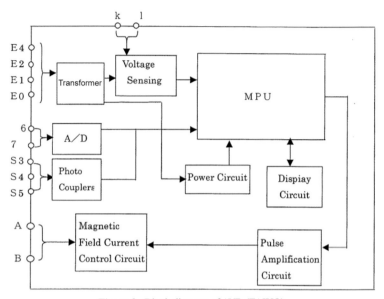

Figure 8. Block diagram of AVR (TAIYO).

4.2 Magnetic Field Current Control Circuits

Figure 9 shows magnetic field current control circuit (Thyristor half-wave shunt circuit). AVR terminals A and B are connected to A.C. side of the rectifier provided in the exciter. A.C. current supplied by reactor and current transformer provided in the exciter is rectified to become magnetic field current. This current is shunted in part by means of phase control of the thyristor provided in the AVR while magnetic field current is regulated.

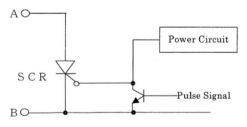

Figure 9. Thyristor half-wave shunt circuit.

4.3 AVR Control Circuits

4.3.1 The circuit principle diagram of TAIYO AVR

TAIYO AVR diagram is shown as Figure 10.

(1) Voltage detection circuit

Generator voltage is reduced by the detection transformer, passed through the voltage detection circuit and RMS detected by MPU. At the same time, a synchronized signal is supplied from this transformer to MPU and the power is supplied to the power circuit.

(2) Droop

The current supplied by current transformer CCT provided in the power distribution board is supplied to terminals K and L, insulated by current transformer DCT provided in the voltage detection circuit, converted into 1/1000 and then converted to a voltage signal.

This signal is proportional to the load current from vector viewpoint. The signal and generator detection voltage are vector composed in the detection circuit in order to provide appropriate droop characteristics to the generator voltage characteristics.

(3) Micro-processor (MPU)

MPU executes RMS calculation of the generator voltage, PID calculation of the reference value and deviation signal, calculation of timing of signal transmitted to thyristor, alteration and storage of PID constants, control of display circuit and the like. Alteration of PID constants is possible even during operation.

(4) Pulse amplification circuit

In the pulse amplification circuit, a fine pulse signal from MPU is passed through the photo-coupler and is insulated, and a pulse is given to the thyristor through the amplification circuit.

(5) Power circuit

The power circuit supplies 5 V power to MPU, pulse amplification circuit, and display circuit.

(6) Voltage setting circuit

Two circuits are available for voltage setting. One is to use external variable resistor and the other is to use increase/decrease contact signals. An analogue voltage-setting signal from the variable resistor is converted into a digital signal by A/D converter and is input to MPU. Further, a signal from the contact signal is

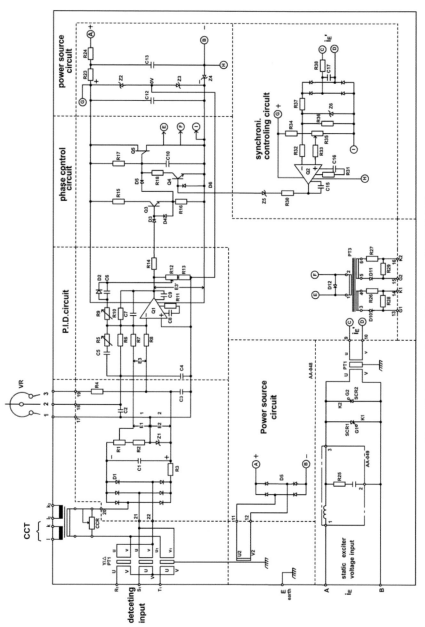

Figure 10. The circuit principle diagram of TAIYO AVR.

capable of transmitting an increase/a decrease signal to MPU via photo-coupler. Thus voltage setting can be changed freely. These signals are set referring to the voltmeter while the generator is being operated.

(7) Display circuit

The display circuit consists of four switches (Enter, Shift, Up and Down push buttons) and numeric display. It is possible to input various constants using these four switches.

4.3.2 The circuit principle diagram of GA2491 AVR

GA2491 AVR diagram is shown as Figure 11.

The generator voltage is fed to the regulator via plug connector X_1 in a single-phase, two-circuit arrangement. Transformer T_1 steps down the generator voltage which is then rectified by the load-side rectifier bridge V_1, V_4. This rectified voltage provides the actual pulse signal "U_{ist}", the setpoint voltage U_{soll}, and the supply voltage for the regulator.

If the system uses a reactive current compensator, current transformer T_{15} or interposing transformer T_4 of the excitation unit is connected to load resistor R_1 via plug-in contacts $X_{2/5}$ and $X_{2/9}$. In this operating mode, the actual voltage is composed of the secondary voltage of transformer T_1 and the voltage of load resistor R_1.

The magnitude of the resulting reduction in generator voltage can be set with potentiometer S. If an external setpoint selector is used, this is connected by contacts $X_{2/1}$ (A_1) and $X_{2/3}$ (A_3). In this case, microswitch $S_{1/3}$ of the regulator must be opened.

A DC voltage of 0~10 V can be fed in via plug-in contacts $X_{2/6}$ and $X_{2/2}$. This voltage acts on the comparator point of the control amplifier. The set point can thus, for instance, be preset by higher level equipments.

Control amplifier ② (proportional again adjustable by potentiometer K and reset time by potentiometer T) outputs a DC voltage which is converted into a time adjustable firing pulse for thyristor V_{18} or V_{28} via the load side pulse unit. The generator excitation circuit is fed from rectifier bridge V_{29}. Resistor R_{48} and thyristor V_{28} form a parallel bypass circuit to the field winding through which part of the current supplied by the excitation unit flows. This method provides for generator voltage control. In order to optimize the correcting action, a disturbance variable is injected into the control amplifier via resistor R_{47}.

Over voltages above DC 600 V in the excitation circuit cause the overvoltage protector to operate and continuously fire the thyristor. Protection is thus provided for the stationary excitation circuit of the generator.

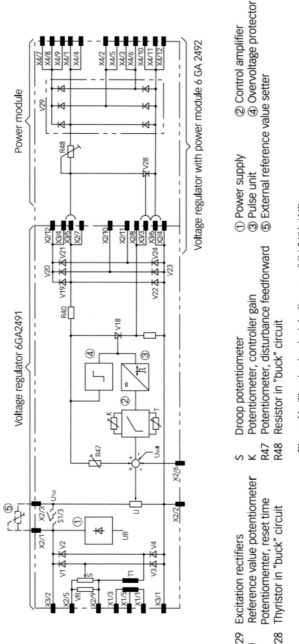

Figure 11. The circuit principle diagram of GA2491 AVR.

V29 Excitation rectifiers
U   Reference value potentiometer
T   Potentiometer, reset time
V28 Thyristor in "buck" circuit

S   Droop potentiometer
K   Potentiometer, controller gain
R47 Potentiometer, disturbance feedforward
R48 Resistor in "buck" circuit

① Power supply   ② Control amplifier
③ Pulse unit   ④ Overvoltage protector
⑤ External reference value setter

196

(1) Thyristor voltage regulators

The voltage regulates the voltage so that it complies with the set point selected. Frequency changes due to the droop characteristics of the prime mover do not affect the voltage accuracy. The design and adjustment of the generators and the excitation equipments permit continuous changes of the terminal voltage in the range of ± 5% rated voltage via the set point selector under steady-state conditions and at loads varying from no load to rated load, and power factors from 0.8 to unity unless specified otherwise on the rating plate.

If several rated voltages and frequencies are indicated on the rating plate, the above data apply to each of the rated voltages stated. If the generators are operated at voltages exceeding ± 5%, the generator output must be reduced. Unrestricted operation at no load is permitted if the speed is reduced.

During operation, the excitation circuit must not be interrupted since this would give rise to voltage surges. If the generator must be de-excited, this can be accomplished by short-circuiting the secondary side of rectifier transformer (T_6).

(2) Transformer adjustment

The tapping used on the transformers is shown in the test report. It is strongly advised not to change the original adjustments. No responsibility can be assumed by the supplier for any damage or incorrect operation resulting from a change in the original adjustments. In the case of identical plants, the THYRIPART excitation system or the individual components can be interchanged if necessary. The transformer tapping, however, must be used in accordance with the original ones.

(3) Regulator gain, setpoint voltage integral action

The control module comprises potentiometers U, K, T, R_{47} and S. The rated generator voltage has been adjusted in the factory on potentiometer U, and the dynamic behavior of the regulator on potentiometers K, T and R_{47}. The settings are shown in the test report. Potentiometer K is used to adjust the controller gain and potentiometer T is used to adjust the integral action time, whereas potentiometer R_{47} is used to inject a disturbance variable into the comparator point of the control amplifier in order to adjust dynamic behaviour. Turning the knob of K and R_{47} in the direction of descending numerals and that of T in the direction of ascending numerals normally stabilizes the control circuit and reduces the control rate. The stability of the control circuit can also be improved by increasing the bucking resistance, but the voltage setting range of the regulator is then reduced at the lower band.

The setpoint of the generator voltage can be shifted via potentiometer U or an additional external setpoint selector (R = 4.7 kΩ, P greater than 1 W) can be connected to terminals A_1 and A_3. Potentiometer U should be set to the centre position, and microswitch $S_{1/3}$ on the printed-circuit board should be opened.

197

(4) Parallel operation by droop compensation equipments

When provided with droop compensation equipments, brushless synchronous generators are suitable for operating in parallel with each other or with a supply system. The KW output is adjusted by the governor of the prime mover. The speed characteristic of the prime mover should be linear and rise by at least 3% and not more than 5% between rated load and no load. Droop compensating equipments ensure uniform distribution of the reactive power and reduce the generator output voltage linearly with the increase in reactive current. Regarding generators with current transformer for droop compensation, potentiometer S in the regulator is adjusted so that there is no reduction in the generator voltage at unity p.f. but a 4% reduction at zero p.f.

The corresponding voltage reduction at 0.8 p.f. is 2.4%. In isolated operation and at any loading condition of the generator, the droop compensation provided for the generator voltage can be checked with the following relationship:

$$\Delta U_{st} = 4\% \sqrt{1 - \cos^2 \phi} \cdot IB / IN (\%)$$

E.g., at 0.8 p.f. IB/IN = 1; $\Delta U_{st} = 4\% \sqrt{1 - 0.8^2} \cdot 1 = 2.4\%$.

If the generators are to operate by itself, droop compensation equipments are not required. They can be deactivated by short-circuiting the secondary side of the associated current transformer or setting potentiometer S on the regulator to the left-hand stop.

Droop characteristic curve is shown as Figure 12.

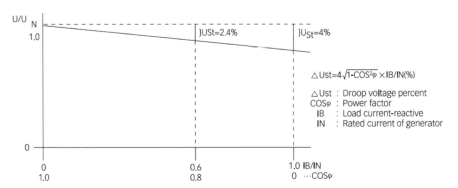

Figure 12. Droop characteristic curve.

(5) Parallel operation by cross-current compensation

When provided with cross-current compensation, brushless synchronous generators are suitable for the operation in parallel with other generators of the same capacity.

This parallel operation by cross-current compensation has the same voltage under all loads condition from no-load to rated load. If the neutral points of several generators are interconnected or connected directly with the neutral points of transformers and loads, currents at 300% frequency may occur. Their magnitude should be checked by measurements in the neutral conductors of the generators under all load occurring conditions. To avoid overheating the generators, these currents must not exceed a value equal to about 50% of the rated generator current. Higher currents should be limited by installing neutral reactors or similar means.

5. Construction and Function of the Static Exciters

The static exciting system is composed of a reactor (RT), current transformer (CT), silicon rectifier (S_i), semiconductor automatic voltage regulator (AVR), etc. Normally, voltage regulator (VR), parts for parallel running are installed in the switchboard as shown in Figure 13, while the other parts are placed in a box (Figure 14) of drip-proof construction mounted on the top of the generator.

Figure 13. Exciting unit mounted on generator TOP.

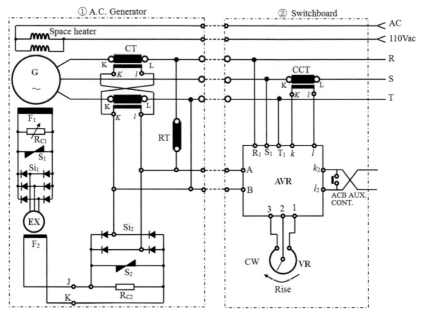

Figure 14. Connection diagram.

5.1 Components of the Static Exciters

(1) Reactors (RT)

The core is provided with clearances in which magnetic insulators are inserted. The role of the reactors in the exciting system is such that the exciting current component required to maintaining the rated voltage, when the generator is under no load, is supplied to the field winding of the generator from the output of the generator through the reactor and the silicon rectifier.

(2) Current transformers (CT)

The current transformers have a core provided with the primary and secondary windings. The role of the current transformer in the exciting system is such that the exciting current component required to compensate for the voltage drop, due to the armature reaction and leakage reactance caused by load current, is supplied to the field winding of the generator from the secondary winding through the silicon rectifier.

(3) Silicon rectifiers and protecting devices (Silistor)

The rectifier element, by nature, tends to pass electric current only in one direction and almost no current in the opposite direction. The silicon rectifiers have a peak reverse voltage and would be damaged by the application of an overvoltage, which exceeds the peak reverse voltage. There, the silicon rectifiers must be used together with protecting devices against overvoltage (surge voltages).

(4) Discharge resistors (protective devices)

Discharge resistors are integrated into static exciters. When an overload is applied to the silicon rectifying modules, the discharge resistors reduce overload by applying current proportional to the voltage to protect the silicon rectifying modules.

(5) Current Transformers for Cross-current Compensation (CCT)

The current transformers for cross-current compensation is integrated into panel. The current transformers for cross-current compensation is double molding type metering of current transformers. They are used in detecting load current at parallel running.

6. Generators Operation

When operating the generators, measure the insulation resistance between each of the windings of the generator and ground with a 500 V megger. It is desirable that the measured resistance is above one MΩ. In order to avoid applying the megger voltage to the AVR, be sure to disconnect or short the terminals prior to measurement.

At the time of testing at factory, adjustments are made in relation to power factor change, rotating speed change, load variation, etc., so that, when the rated rotating speed is approached, the voltage will build up within ± 1.0% of the rated voltage. Therefore, there is no need to control the voltage regulator.

6.1 Parallel Running

For parallel running of generators, refer to the following operating principles and procedures. Erroneous handling may cause a generator failure. In order to operate a generator in parallel with another generator in operation, it is necessary for both generators to satisfy the following conditions (Zhang 2013):

To be identical in frequency
To be equal in voltage magnitude
To be matched in phase

Conditions required of the prime movers
To have uniform angular velocities
To have an adequate speed regulation

In case the above mentioned conditions are not satisfied, the following undesirable phenomena and situations will take place. In case there is no matching in phase, much crosscurrent will flow, and closing for parallel running will be impossible. Any difference in voltage magnitude will cause a flow of crosscurrent between the generators, so that the generators with higher and lower voltages will have lagging and leading currents respectively, and thus reactive power shedding will be impossible (Huang and Wang 2007). If the generator load's characteristics do not include a drooping characteristic with respect to reactive power, stable shedding of reactive power is impossible and this may give rise to hunting or step-out.

In case the angular speed is not uniform, load shedding cannot be performed. The prime mover is provided with an adequately sized flywheel for angular velocity uniformity. Without governor characteristics, load shedding is not stable (Zhu and Sun 2006). If the governor speed's characteristics do not include stability or quick response, just as in the case of governor failure load shedding is unstable, or hunting or step-out may result.

6.2 Synchronizing

Carry out governor adjustment on the prime mover side until the frequency of the generator is to be connected in parallel with the forerunning generator is approached. While watching the synchronizing lamps or synchroscope, finely adjust the governor, and close the circuit breaker at the point of synchronism. As to the synchronizing lamps, the time when one of the lamps has gone off and the two others have become brightest is the phase coincidence point. As to the synchroscope, the time when the needle has come to the top center is the phase coincidence point (Lv et al. 2008). It is a proper method to start the closing operation with the closing time of the circuit breaker taken into account, and to close the circuit breaker just at the point of synchronism (when the needle of the synchroscope has come to the top center).

6.3 Load Shedding

After completion of synchronizing, while paying attention to frequency, operate the governor on the forerunning generator's side in the speed lowering direction and operate the governor on the after-running generator's side in the speed raising direction (Zhang and Ma 2011). By doing so, make a gradual load shift for load shedding.

6.4 Release from Parallel Operation

When disconnecting one of the generators running in parallel from the bus, take the following steps: operate the governor of the driver for the generator to be released in the speed lowering direction and operate the governor of the driver for the generator on the bus side in the speed raising direction for load shifting. When the load on the generator to be released becomes zero, trip the circuit breaker for the same generator.

7. Maintenance

7.1 Temperature Rise

The monitoring of the temperatures of devices is an effective means for finding faults, if any. Check temperature rise in components by referring to Table 1.

Check the bearings on a daily basis, and pay attention to any rapid temperature rise. The maximum bearing temperature during normal operating conditions is on the order of 80–85°C (thermometer reading).

Table 1. Allowable Temperature Rise Limits (°C) (Standard ambient temperature: 50°C).

Machine part	Measuring method	Allowable temperature rise limit	
		Insulation class B	Insulation class F
Armature winding	By thermometer	60	75
(Generator stator winding)	By resistance	70	90
Field winding of cylindrical rotor	By resistance	80	100
Core or other machine parts adjacent to insulated winding	By thermometer	70	90
Bearing	When measured by temperature detector element embedded in metal: 35		

7.2 The Replacing of the Bearing

To check the clearance of the bearing, measure the outside diameter of the shaft and the inside diameter of the bearing at several points by micrometer. If it exceeds the bearing clearance calculated by the following formula, replace the bearing.

$Y = (1.5 \cdot X)/1000 + 0.1$

Where Y-bearing upside clearance (mm); X-Shaft diameter (mm).

7.3 Judgment of Measured Resistance

The resistance measured by 500 V megger is acceptable if it is above the value calculated by the following formula:

Insulation resistance = (3 × Measured voltage (V))/(Rated output (KW or KVA) + 1000) (MΩ)

The best way to judge the insulation is to compare the newly measured resistance with the value obtained by previous measurement. The variation from the preceding measurement result is significant so that insulation resistance measurement should always be continually carried out with recording.

7.4 Check the Silicon Rectifiers

The Silicon Rectifiers may become faulty if an overvoltage is applied or if an over-current flows. When the silicon rectifiers becomes faulty, almost no generator voltage will be produced. For simply judging the silicon rectifiers to be good or not, measure resistances using a tester by the procedure illustrated in Figure 15. Disconnect all the wires connected to the silicon rectifier stack. Using the resistance measuring range of the tester, measure the forward and reverse resistances of each rectifier element.

The forward resistance is acceptable if it is below 10Ω, while the reverse resistance is acceptable if above 100 kΩ. Any rectifier element that is conductive

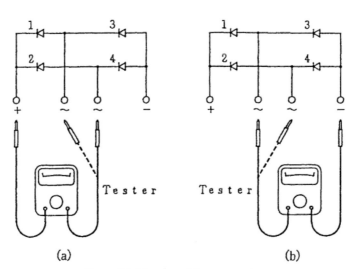

Figure 15. Measure resistances procedure.

in both forward and reverse directions is defective. If a faulty element has been found by the above measurement, replace it with a good element.

7.5 Check the Rotary Rectifiers

The rotary rectifiers exhibit stable performance with excellent mechanical and electrical characteristics. So, correct operation and maintenance will allow the rotary rectifiers to be used without hindrance for a long time. The rotary rectifier is incorporated in a three-phase full-wave rectifying circuit as shown in Figure 16. As shown, a Si varistor (Silistor) is connected for sure protection of the silicon rectifiers.

The construction of the rotary rectifiers vary according to generator's model and A.C. exciter incorporating method.

It is important to keep the rotary rectifiers and the surroundings clean just like the stator and the rotor windings of the main generators and exciters. Carry out

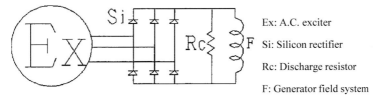

Ex: A.C. exciter

Si: Silicon rectifier

Rc: Discharge resistor

F: Generator field system

Figure 16. Rectifying circuit.

cleaning periodically just as for the windings. Also, at the same time as cleaning, check for any grounded terminal or lead wire and any loose bolt or nut around the rotary rectifier.

8. Conclusions

Marine brushless A.C. generators for synchronous machines have been introduced in this work. The construction of the marine generators and principles of exciting system have been analyzed. The experience in operation of marine generators has been proposed in this paper. Merchant ships will benefit greatly from improvements in generators and control in the marine power systems. Maintenance experience of marine generators can be used as a reference for marine engineers on board. The current developments will improve their power management ability.

References

Guichen Zhang and Jie Ma. (2011). Speed-torque hybrid control with self-tuning for marine power station. Journal of Shenyang University of Technology, 33: 434–438.

Guichen Zhang. (2013). Modern Marine Power System. Dalian Maritime University Press. Dalian. China, pp. 75–83.

Jinfan Jiang. (2005). Ship Power Station and Automation. Dalian Maritime University Press. Dalian. China, pp. 69–75.

Man-lei Huang and Chang-hong Wang. (2006). Nonlinear mathematical model of diesel-generator set on ship. Journal of Harbin Engineering University, 27: 15–19.

Man-lei Huang and Chang-hong Wang. (2007). Research on double-pulse H-infinity speed governor for diesel engine of ship power station. Control Theory & Application, 24: 283–288.

MV COSCO BELGIUM. (2013a). High Voltage Main Switchboard Instruction. Shanghai. Open Shipping Co. Ltd., China.

MV COSCO BELGIUM. (2013b). Main Generator Instruction. Open Shipping Co. Ltd., Shanghai. China.

MV COSCO XIAMEN. (2004). Low Voltage Main Switchboard Instruction. Open Shipping Co. Ltd., Shanghai. China.

Qi-weiLv, Dian-li Zhao and Chun-lai Zhang. (2008). Implement of automatic quasi-synchronization of marine power station based on PLC. Journal of Dalian Maritime University, 34: 76–78.

Shilong Xue. (2011). Ship Power System and Automatic Control, Beijing. China.

Yang Zhang and Yi Wu. (2013). Design of synchronous generator for integrated power propulsion system in marine. Transactions of China Electrotechnical Society, 28: 67–74.

Zhi-yu Zhu and Chuan Sun. (2006). Robust controller design of ship decentralized excitation generator. Electric Machines and Control, 10: 474–480.

9

Marine Main Switchboards

Guichen Zhang[1] and *Zenghua Sun*[2]

1. Introduction

Marine low voltage and high voltage main switchboard are designed for carrying out the control of single run or parallel run of the AC generators according to demand for the electric load on board. It is are assembly of switching controls, regulating devices, measuring instruments, indicator lamps, and protective devices and distributes electric power to the various loads on board (Zhang 2013). These equipments completely protect the electric equipments by respective systems in case of electric accident. They are dead-front type, is sturdy as constructed with steel plates and angle bars, is designed for easy inspection and maintenance. The inspection and maintenance of the instruments inside can be easily carried out by opening the front door or rear cover (MV COSCO BELGIUM 2013).

The faults comprise incorrect indication of various meters, non-indication of pilot lamps, non-working or mis-working of circuit breakers or magnetic contactors. Their probable direct causes are: breaking of circuit, grounding of circuit, short-circuiting, imperfect contact of circuit, loose fastening of connecting screws, etc. The High voltage switchboard are equipped with the circuit breakers which protect the generators from faults. They not only feed the electric power to the loads, but also protect the load side equipments correctly responding to the kind of faults

[1] School of Naval Architecture and Ocean Engineering, Harbin Institute of Technology, No. 2, West Wenhua Road, High-tech District, Weihai, Shandong, P. R. China.

[2] Marine Engineering College, Dalian Maritime University, No. 1 Linghai Road, Dalian, Liaoning, P. R. China.

(Jiang 2005). Generators are normally controlled in automatic or semi-automatic mode. Even if generators are out of control in automatic or semi-automatic mode, they can be controlled in manual control mode (Xue 2011).

2. Marine Low Voltage Main Switchboards (LVMSB)

Generator remote control panel (MV COSCO XIAMEN 2004) is shown as Figure 1; The Parallel switchboard (MV COSCO XIAMEN 2004) is shown as Figure 2.

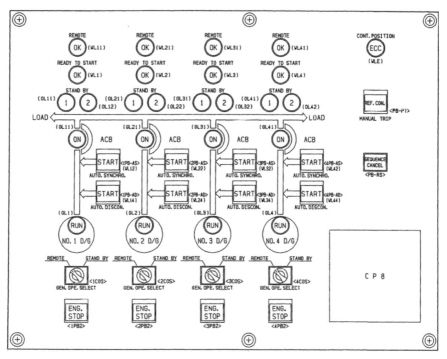

Figure 1. Generator remote control panel.

2.1 Running and Stopping Diesel Generator

2.1.1 Single running

(1) Confirmation items for the preparation of operation

When the rating value is not clear, set the value at minimum and increase the voltage gradually to the rated value after starting the engine. Air circuit breaker is in open condition; voltage regulator is in rating value.

IL1: DC24V Source (WL)
IL2: Spare
IL3: Ready to start (WL)
IL4: 1ST ST-BY (WL)
IL5: 2ND ST-BY (WL)
IL15: E/G Run (GL)
IL16: E/G Auto ST-BY (WL)
PL1: Auto Synchro Start
(Push button with WL)
PL2: Auto Load Shift Start
(Push button with WL)
FM: Frequency Meter
TL: Transparency Lamp
SY: Synchro Scope
SYL: Synchro Lamp
SYS: Synchro Scope
COSP: Synchro & Load Sharing
BCS: Breaker Control Switch
GCS: Governor Control Switch
CSE: ENG. Control Switch
SW21: ST-BY GEN Select
VS1: VOLT Meter (Bus)
VS2: VOLT Meter (GEN & Shore)
FVS: FREQ & VOLT Meter
PB1: Lamp Test
CNP: Caution Name Plate
NP : Name Plate

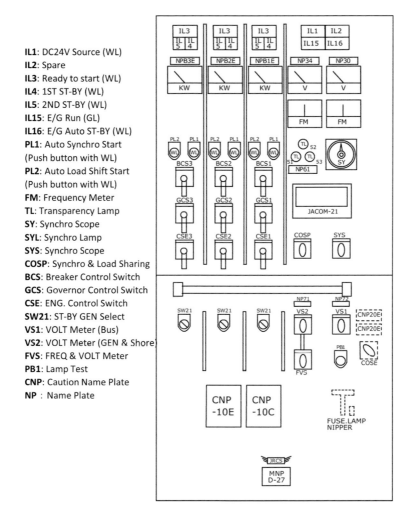

Figure 2. Parallel switchboard.

(2) Start the generator diesel engine

On starting the diesel engine, generator voltage is increased followed by the increase in engine's speed and lightning of the power source lamp. The starting the diesel engine, generator voltage increase followed by the increase in engine speed and power source lamp switches on. The procedure to start the diesel generator is shown in Figure 3. If the diesel engine does not reach ignition speed, the chosen restart program will be repeated N-times (N < 3) after a breaktime. Supervise only in automatic mode.

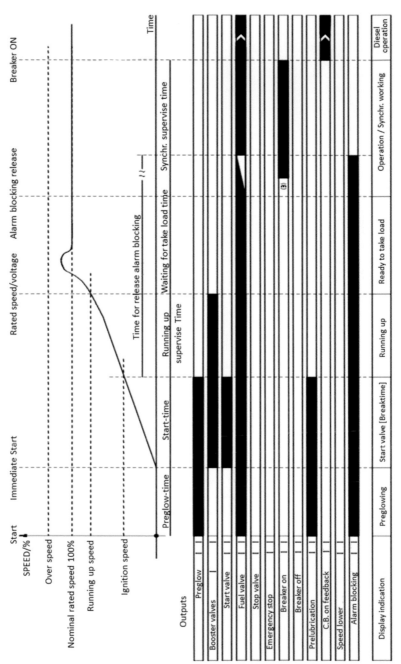

Figure 3. Starting procedure of the diesel generator.

(3) Confirmation and adjustment of voltage and frequency

> a. By operating the voltage meter switch, please check the voltage between the phases of generator and adjust the voltage to the rating value by voltage regulator. In this case, voltage rises by turning the voltage regulator clockwise and voltage decreases by reverse operation.
>
> b. After voltage is set to the rating value, measure the frequency by operating the frequency meter switch. When the frequency is not the same as of the rating value, adjust the value by the governor switch. The Governor switch returns to "OFF" position automatically by releasing it.

(4) Preparation of air circuit breaker closing is finished with the above described operation, but the closing operation must be carried out after change over the voltage meter switch and it should be confirmed that the bus bar is in non-voltage condition. ACB (Air Circuit Breaker) closed lamp is on condition.

(5) Power is supplied on the main switchboard bus bar by the above operation and then the moulded circuit breaker of the demanded load is closed.

2.1.2 Parallel running of generators

When the demanded load on board increases or when the change in generators from running to stand-by is to be carried out, the following conditions are to be satisfied before synchronizing the generator (Zhang and Ma 2011).

> a. Frequency difference between generator and bus bar is small enough;
> b. Voltage difference between generator and bus bar is small enough;
> c. Phase difference between generator and bus bar is small enough.

When the synchronizing is manually carried out, the above condition must be satisfied by manual operations. The following steps explain the method of synchronizing between stand-by generator and the running generator (Zhang et al. 2016).

(1) Stand-by generator is started by the method described (2.1) and adjusts the voltage to the rating value.

(2) Confirmation and adjustment of Voltage and Frequency

> a. By operating the voltage meter switch, confirm that the each phase voltage is equal to the bus voltage.
> b. By operating the frequency meter switch, confirm that the frequency is equal to the bus frequency.

(3) Confirm the pointer hand of synchronizing meter by changing the synchronizing switch to the stand-by generator side. At this time, the synchronizing lamp lights on and off t same time.

    a. When the frequency difference between the generators is higher than 3Hz, only synchronizing meter lamp lights off. The frequency difference has to be reduced by the governor switch.

    b. When the synchronizing meter rotates to the "FAST" side, this shows that the frequency of closing generator is higher than bus bar frequency.

        In this case, change over the closing generator engine governor switch to the "SLOW" side and adjust the frequency

    c. On the contrary, when the synchronizing meter rotates to the "SLOW" side, this shows that the frequency of closing generator is lower than the bus bar frequency.

        In this case, change over the closing generator engine governor switch to the "FAST" side and adjust the frequency

    d. When the synchronizing meter stops at the top (synchronized point), it shows that both generator phase and frequency has matched each other.

(4) After confirming matching through the synchronizing meter, close the air circuit breaker. After having finished the closing operation of air circuit breaker, change over the synchronizing switch to the "OFF" position.

For the synchronizing operation, phase matching operation is the most difficult one, especially it is very hard to stop the synchronizing meter on the top place. Then, adjust the synchronizing meter pointer rotation very slowly by the governor switch (Sheng et al. 2017).

(Difference is within 0.15 Hz): In this case, the rotating direction is nothing with "SLOW" or "FAST" side, but when the synchronizing operation is carried out at "SLOW" side rotating, pay attention to the air circuit breaker trip since the relay is energized by reverse power before the operation of load sharing. For the closing operation, it is important to start the closing operation with the consideration of closing time before the synchronizing meter pointer goes on the top (synchronized point). Take care of the miss operation because the closing misoperation at synchronizing causes blackout, i.e., mechanical damage or electrical damage to the generator and may also induce a serious trouble on board.

(5) Load sharing

Generator synchronizing operations end with the above mentioned operations. Next, is the load sharing operation.

a. When closing the generator engine, governor switch is operated to the "FAST" side and engine revolutions run higher. The working generator load is shifted to the closing generator in a little capacity and the previous working generator load is reduced and both frequency rise.

To prevent these conditions, the previous working generator's engine governor switch is operated to "SLOW" side and previous working generator's engine goes down. The previous working generator's engine load is reduced but closing generator load will increase the load. Adjust both the watt meter to make the same value appear by adjusting both the governor switches.

2.1.3 Load shifting

Adopt the following method when one set of engine is stopped at parallel running condition. This explains the method to stop the one set of running engine at parallel condition.

a. The diesel engine which is going to stop, its governor switch is set to the "SLOW" side, and at the same time, the diesel engine which is going to continue to run, its governor switch is set to "FAST" side then.
The entire loads are shifted to the continuous running generator's side. When the power meter of the generator which is ready to stop indicates "0", immediately switch off the air circuit breaker which is going stop. These operations must be carried out with checking the frequency.
b. After air circuit breaker is changed over to "OFF" position, diesel engine is stopped.

2.1.4 Stopping generator

When the generator is going to stop, keep the operation order as follows, never carried out in opposite operation order. Stopping procedure of the diesel generator is shown in Figure 4. ACB OFF with load reduction is active as long as the breaker is switched off. Event is active as long as the breaker is closed (Lu et al. 2017).

① Take off the load completely;
② Turn to the "OFF" position of the air circuit breaker;
③ Engine is stopped at engine side.

2.2 Shore Source Interlock

(1) This explains the source connecting from shore connecting box to main switchboard

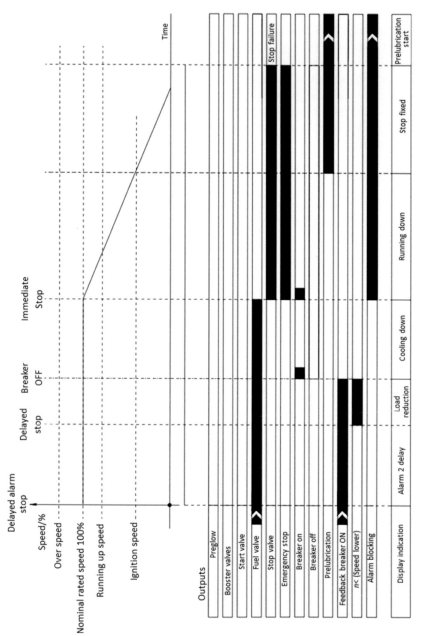

Figure 4. Procedure to stop of the diesel generator.

a. Wire the cables in normal phase revolution to the shore connecting box and close the molded circuit breaker (M.C.B.).

b. Confirm the voltage by operating the voltage meter switch on the main switchboard.

c. After that, switch off all the air circuit breakers. Power source is supplied from shore source by switching on the shore source MCB.

(2) This explains the interlock between MCB of shore source (herein after SCMCB) and generator's ACB (hereinafter ACB).

a. When one of the ACB is closed condition, the circuit of the under voltage trip mean (UVT) in the SCMCB is broken by auxiliary connector in the ACB, then SCMCB cannot be closed by the operation of UVT.

b. On the contrary, when the SCMCB is in the closed condition, auxiliary connector breaks the UVT circuit in the ACB through the auxiliary relay, then ACB cannot be closed.

c. Interlock is provided as in the above description but beware of the operation.

2.3 Space Heater Interlock for the Generators

Space heaters are used to heat with heater to prevent the decrease of generator winding insulation resistance by the moisture while the generator is in the stop condition. The space heater interlocks with the generator's ACB (Automatic Air Circuit Breaker), if the ACB is closed unless the space heater switch is "OFF" condition, ACB auxiliary connector energized and cut the space heater circuit to prevent the excess heating.

2.4 Earthing Detector Lamp

(1) When the line test switch is in the "EARTH MONITOR" position, three earthing detector lamps light with same brightness. Earthing detector lamp electrical circuit is shown as Figure 5.

(2) Next, earthing condition of each phase can be checked by changing over the line test switch to the "EARTH TEST" position.

a. If someplace of R-phase is in incomplete earthing condition, earthing detector lamp of R-phase goes dark and does not light on in case of complete earthing.

b. If all the phases are not in earthing condition, three lamps light with the same brightness.

c. Return the line test switch to "EARTH MONITOR" position after finishing the earthing test.

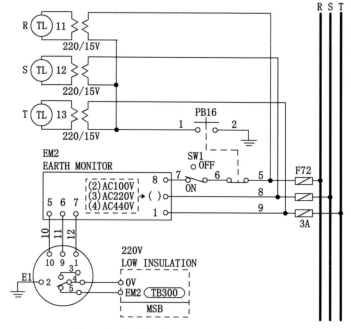

Figure 5. Earthing detector lamp electrical line.

2.5 Preferential Trip Circuits

When the running generators become overload, predetermined electric load MCB are tripped preferentially to decrease the running generator load and continue to run (Xie et al. 2017). Shunt trip coil in the MCB is energized by the output signal of the over current sensor and electric load is tripped.

2.6 Insulation Resistance Meters

By the changeover of line test to "EARTH MONITOR", results in change in resistance meters on the switchboards. Insulation resistance meter's electrical line is shown as Figure 6.

3. Marine High Voltage Main Switchboard (HVMSB)

3.1 Major Components of HVMSB

HVMSB (MV COSCO BELGIUM 2013b) is shown as Figure 7.

(1) Vacuum Circuit Breakers (VCB)

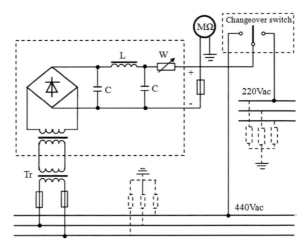

Figure 6. Insulation resistance meter electrical line.

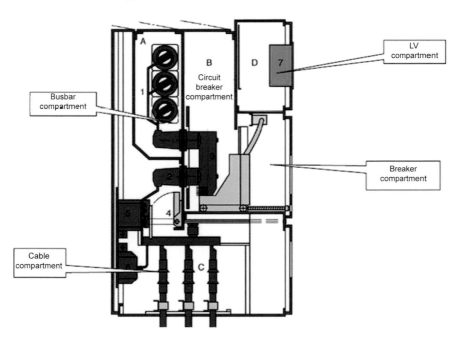

Figure 7. High Voltage Main Switchboard.

The circuit breakers are of the vacuum type (VCB), draw-out type with automatic shutter and earthing switch. They are operated by an electrically operated, mechanically and electrically trip-free, motor spring charge operating mechanism.

217

The provision is included for manual operating of the mechanism. The circuit breaker compartments are furnished with mechanism which will move the breaker between the operating and disconnecting positions (MV COSCO BELGIUM 2013b). Control voltage of the circuit breaker is 110 V DC from the outside of HVMSB.

(2) Vacuum Combination Contactor (VCCT)

The contactors are of the vacuum type (VCC), they are used for the motor main circuit and are of draw-out type with fuse. The contactors are operated by an electrically operated. Control voltage of the contactor is 110 V DC from the outside of HVMSB.

(3) Multiprotection and Control Unit for Generator (HIMAP-BCG)

The multi-protection and Control unit HIMAP-BCG is a control unit supplying power management system. This is used for generator protection and protection functions are based on the IEC 60255. This is installed on each generator panel front (Ge et al. 2017).

(4) Multi-functional Digital Protection Relay for Feeder (HIMAP-FI)

The Multi-functional digital protection relay HIMAP-FI is a protection relay unit and communication facilities are provided to enable measuring functions to be monitored remotely and power management system to operate automatically. This is used to protect feeder circuits, installed on each feeder panel front (Liu et al. 2017).

(5) Generator Plant Management System (GAC)

The "Generator Plant Management System" always monitors both the data transmission system and the operation of the GAC controller. The GAC controls generating plant separately so that extremely high level of power supply reliability is achieved.

(6) Earthing Switches (ES)

Earthing switches are of fixed type and are installed in the medium voltage vacuum circuit breaker and vacuum combination contactors unit on the under side. Mechanical interlocks are provided between the vacuum circuit breakers and the earthing switches.

(7) Interlock Arrangement

Mutual interlocks are provided among the generators VCB or ACB, bus-tie VCB or ACB respectively equipped with High Voltage Main Switchboard (HVMSB),

Low Voltage Switchboard (LVSB), No. 1 440 V Service Transformer Panel (TR1), and No. 2 440 V Service Transformer Panel (TR2). This aims to maintain optional status of power supply through opening or closing of each VCB or ACB, when normal power supply is disturbed by some reason or depending on the relation between the number of generators on-line and power received or the demand-supply situations of power. Opening/closing operation of the VCB or ACB is done by electrical closing operation or mechanical closing operation depending on the application purpose.

For safe maintenance and checkup of the compartment which accommodates each component of the HVMSB, each cell is de-energized by opening the main circuit breaker and power line charging current is discharged by closing (earthing) the earthing switch. To do this, therefore, the interlock by key operation, according to the predetermined operating procedure, is provided between the main circuit breakers, earthing switches, and compartment rear door.

(8) Earth Lamp

Figure 8 is a simplified circuit diagram to show ground fault detect arrangement. The "EARTH LAMP" (EL) consists of a display lamp having three transparent lenses.

Three earth lamps are openly connected among the phases. Each earth lamp is turned on with the same degree of brightness. When "GROUND FAULT" occurs.

If ground fault occurs in R-phase, the brightness of earth lamp for R-phase becomes lower, while the brightness of earth lamps for S-phase and T-phase becomes higher as compared with that of R-phase. This indicates ground fault of R-phase. When R-phase is completely in the ground fault state, the earth lamp of R-phase is turned off (Yuan et al. 2017). When there is no ground fault in any of these phases, three earth lamps continue to show the same degree of brightness.

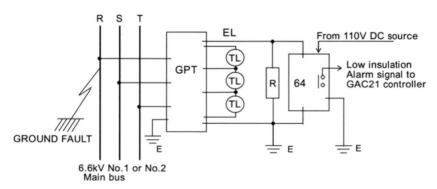

Figure 8. Ground Fault Detect Circuit.

4. Conclusions

LVMSB and HVMSB have been introduced in this work. The construction and principles of MSB have been analyzed. The experience in operation of MSB has been explained in this paper. Merchant ships will benefit greatly from improvements in MSB. Current development will result in MSBs that significantly reduce the cost and losses of power. With the application of advanced technology, developmental trends will focus on advanced MSBs, standardization and networking.

References

Bo Yuan, Hua Shao, Chunguang He, Ying Wang, Ting Yang and Ying Wang. (2017). Key issues and research prospects of smart distribution system planning. Electric Power Automation Equipment, vol. 37, No. 1, pp. 65–73.

Guichen Zhang and Jie Ma. (2011). Speed-torque hybrid control with self-tuning for marine power station. Journal of Shenyang University of Technology, vol. 33, No. 4, pp. 434–438.

Guichen Zhang. (2013). Modern Marine Power System. Dalian Maritime University Press. China, pp. 18–39.

Hong Liu, Jifeng Li, Jiaan Zhang, Hao Sun, Wei Liu and Gaoqing Qu. (2017). Power supply capability evaluation of medium voltage distribution system considering reliability. Automation of Electric Power System, vol. 41, No. 12, pp. 154–160.

Jinfan Jiang. (2005). Ship Power Station and Automation. Dalian Maritime University Press. China, pp. 9–12.

Min Xie, Xiang Ji, Sahojia Ke and Mingbo Liu. (2017). Autonomous optimized economic dispatch of active distribution power system with multi-microgrids based on analytical target cascading theory. Proceedings of the CSEE, vol. 37, No. 17, pp. 4911–4921.

MV COSCO BELGIUM. (2013a). Main Generator Instruction. Shanghai. Ocean Shipping Co., Ltd., China.

MV COSCO BELGIUM. (2013b). High Voltage Main Switchboard Instruction. Shanghai. Ocean Shipping Co., Ltd., China.

MV COSCO XIAMEN. (2004). Low Voltage Main Switchboard Instruction. Shanghai. Ocean Shipping Co., Ltd., China.

Shaoyun Ge, Han Zhao, Xianrong Lei, Hong Liu, Henghui Lian and Qiyuan Cai. (2017). Evaluation and analysis of load supply capability of high-medium voltage distribution system. Journal of Tianjin University (Science and Technology), vol. 50, No. 7, pp. 639–747.

Shilong Xue. (2011). Ship Power System and Automatic Control, Publishing House of Electronic Industry, Beijing. China, pp. 67–69.

Wanxing Sheng, Qing Duan, Xiaoli Meng, Haitao Liu and Changkai Shi. (2017). Research on the AC&DC Seamless-hybrid fluent power distribution system following the power electronics evolution. Proceedings of the CSEE, vol. 37, No. 7, pp. 1877–1888.

Wei Zhang, Weifeng Shi and Hongqian Hu. (2016). Research on agent based reconfiguration and its optimization for shipboard zonal power systems. Power System Protection and Control, vol. 44, No. 4, pp. 9–15.

Zhigang Lu, Guihong Yan, Liye Ma, Kai Guo, Xueping Li and Hao Zhao. (2017). Security classification in the active distribution system based on the total supply capacity. Proceedings of the CSEE, vol. 37, No. 9, pp. 2539–2550.

10

Generator Automatic Control System

Guichen Zhang[1] and *Zenghua Sun*[2]

1. Introduction

The generator automatic control (GAC) function, one of the main functions of the generating plant management system (hereafter referred to as "GAC system"), works with various input signals for generators which are sent by digital multiplex transmission or as individual signals. GAC functions include generator voltage monitoring, bus voltage and frequency monitoring, generator synchronizing (paralleling), load sharing control, and load shift control (Jiang 2005). The GAC system is composed of the two sets of programmable controller and one set of the digital synchronizer and one set of check synchronizer CSQ-3.

The GAC controller is responsible for the overall management of the generating plant. It has an automatic synchronizing/automatic load sharing function, automatic generator start and changeover commands, and power management functions (Xue 2011). It is used to automate plants with unattended machinery rooms and performs efficient power management for super-rationalization plants. It outputs the necessary control and status signals to the outside of the GAC system by the digital multiplex transmission or as individual signals (MV COSCO BELGIUM 2013a).

[1] Merchant Marine College, Shanghai Maritime University, No. 1550 Haigang Avenue, Pudong New Area, Shanghai, P. R. China.

[2] Marine Engineering Colleg, Dalian Maritime University, No. 1 Linghai Road, Dalian, Liaoning, P. R. China.

The above control functions are roughly classified into the following 3 groups:

(1) Remote Semi-automatic Control Functions and Monitoring Functions

Include general semi-automatic operation control functions for generating plants and various kinds of monitoring functions, error detector functions, and warning indicator output functions that are closely related to the automatic control functions.

(2) GAC-UMS Control Functions

Include the automatic control functions such as automatic start control of stand-by generators, intended for "continuity of service" as required by the ship classification societies for ships with unattended machinery spaces.

(3) GAC-PMS Control Functions

Include the power management functions intended for "fuel-consumption saving" and "efficient management of source and load equipment".

2. Functional Elements of GAC System

One GAC can control a maximum of 5 generators.

2.1 Functional Element

 (1) Control mode selection (MANU-AUTO)
 (2) ACB Close/Open
 (3) Governor motor control (Frequency/Output control)
 (4) Frequency Constant Control in Generator Single Operation
 (5) Engine Start/Stop Control
 (6) Engine Abnormal Trip
 (7) Generator Voltage Detection (Establishment)
 (8) Generator Active Power Detection
 (9) ACB Abnormal Trip Detection
(10) ACB Non-Close Detection
(11) Proportional Load Sharing Control/Frequency Control in Generator
(12) Parallel Operation
(13) Preferential Trip (ACB Abnormal Trip)
(14) Automatic Synchronizing Control/Synchronizing Fail
(15) Automatic Load Shift and ACB Trip Control
(16) Automatic Standby Start and ACB Close
(17) Automatic Start/Stop Order Selection
(18) Number of Generator Running Control

(19) Optimum Load Sharing Control
(20) Large Motor Start Block Control
(21) Bus Voltage Monitor (High/Low)
(22) Bus Frequency Monitor (High/Low)
(23) Alarm Indicator Lamp Output for Generator
(24) Indicator Lamp Output for Generator
(25) Alarm Indicator Lamp Output for Common Use
(26) Indicator Lamp for Common Use
(27) Total Power, Surplus Power Monitor, and Detection
(28) Setting Device Operation
(29) Communication Function
(30) Self-diagnostic Functions

2.2 GAC System Types

There are various types of GAC system, as shown in Figures 1, 2, 3, 4, and 5, GAC includes 4 kinds of PMS, such as centralized, distributed, hybrid type, and microcomputer type (Zhang 2013), and the typical GAC are TAIYO, DEIF and TERASAKI, and so on.

3. Remote Manual Control Functions

3.1 Diesel Engine Start/Stop Control and Abnormal Trip

The "ENGINE CONTROL" switch (ECS) is used to start and stop the generator diesel engine. The GAC Controller does not execute the engine control. The engine control performs other engine control modules, etc. The references mentioned below are some of the examples (MV COSCO XIAMEN 2004, MV COSCO BELGIUM 2013b).

(1) Engine control system

The engine control system is assumed to control engines from major manufactures which are of compressed air starting type or electrical starting type. The following shows the compressed air starting engine control system.

(2) General specifications of the engine control system
① Input signals:
· Start interlock signals, including "turning bar in place" signal, "handle switch operation position" signal, etc. (There are three kinds of signals).
· Local/Remote selector switch.
· Engine Start command signal.
· Engine Stop command signal.
· Engine Speed switch (flame speed).

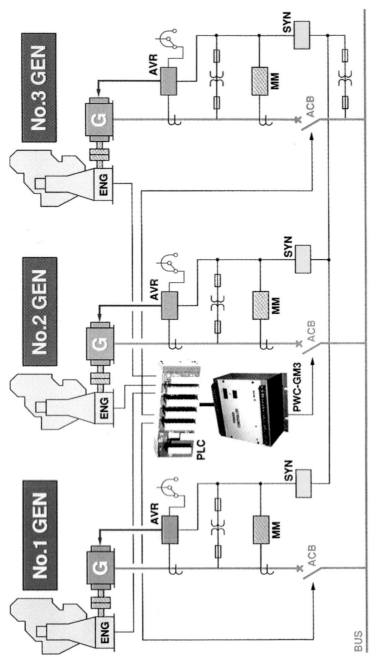

Figure 1. GAC centralized control PMS (TAIYO PWC-GM).

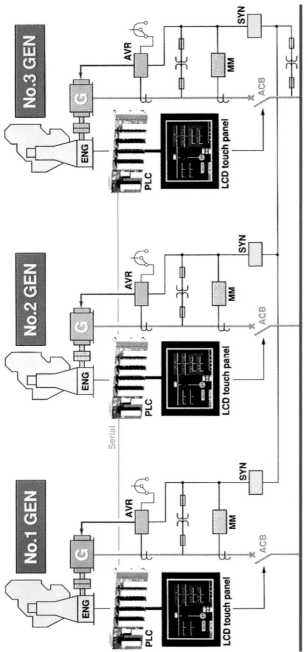

Figure 2. GAC distribution control PMS (TAIYO PWC-PL).

225

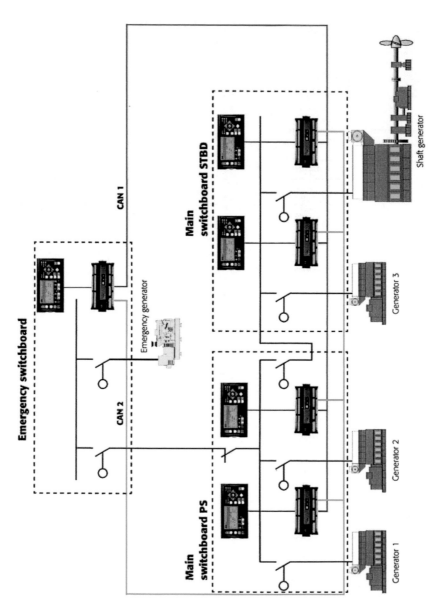

Figure 3. GAC distribution control PMS (DIEF GPU).

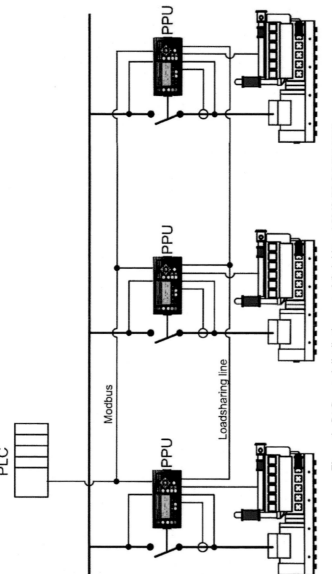

Figure 4. GAC central-distribution control (hybrid type) PMS (PLC-DIEF PPU).

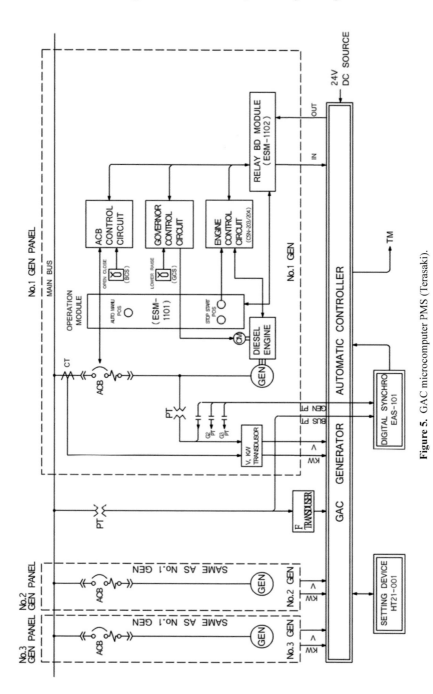

Figure 5. GAC microcomputer PMS (Terasaki).

- Engine Abnormal signals, including "abnormal drop in lubrication oil pressure" signal, over speed signal, and "abnormal rise in cooling water temperature" signal.
- Engine Trouble Reset signal.

② Output signals:
- Start-Valve Operation signal.
- Stop-Valve Operation signal.
- Emergency-Stop Valve Operation signal.
- Abnormal Stop Indicator Lamp signal (Group or individual indication).
- Alarm Repose signal.

③ Control timer:
- Start Fail Detection timer.
- Alarm Rest timer.
- Stop-Valve Excitation timer.
- "Drop in Lubrication Oil Pressure during Start" Tripping timer.

④ Control method of start valve:
Compressed air starting at a specified point in time.

⑤ Control method of stop valve and emergency stop valve:
It depends on the specifications of engines used. An excessive drop in lubricating oil pressure or a rise in cooling water temperature, over speed or the like trip logic function can stop the engine.

3.2 Generator Air Circuit Breaker (ACB) Close/open Control

The "ACB CONTROL" switch (BCS) can be used to close or open the ACB, the GAC Controller has no relation to ACB manual control (Zhang and Ma 2011).

① Manual control
Synchronous closing operation proceeds in manual control mode if a GAC system becomes inoperative due to a trouble of GAC Controller. At that time, manual synchronous closing operation by the SYNCHROSCOPE and "SYNCHROSCOPE" switch (SYS) is performed.

② Semi-automatic control
If required, separate "AUTO SYNCHRO" push-button switch (ASS) will be provided for each generator. The use of this switch allows for activating the synchronous closing function, one of the components for the GAC System, to perform the automatic synchronous closing of ACBs.

Interlock between the generator ACB and shore power connection circuit breaker is provided by hard-wiring logic with the GAC Controller being bypassed.

4. GAC-UMS Control Functions

4.1 Generators Monitoring

The GAC Controller identifies the controlled status of each generator through the following monitoring.

4.1.1 Generator voltage monitoring

The voltage of each generator is monitored to produce the following signals:

(1) Voltage Established Signal

This signal is produced with a time-delay when the generator builds up a required voltage after being started.
- Operate point (VOLT ESTABLISH): 60 to 100% of rated voltage.
- Time delay (VOLT EST TIME): 0 to 30 seconds.
- Reset hysteresis: 20% of rated voltage (Reset point 20% below operate point).

(2) Under voltage Signal

This signal is produced with a time-delay when the generator's voltage has been below a predetermined value.
- Operate point (UNDER VOLT): 60 to 100% of rated voltage.
- Time delay (UNDER VOLT TIME): 0 to 30 seconds.
- Reset hysteresis: 2% of rated voltage (reset point 2% above operate point).

4.1.2 Detection of trouble or delay in generator system control

Here, a "generator system" refers to a generator and its associated equipments under the control of the GAC system. In the event of a trouble or delay in control of a generator system, as listed below, the GAC system will output the associated signal to an external device or system (Wang et al. 2012). If it is a diesel generator system, then the GAC system will disable it from automatically starting and remove it from the stand-by generator list.

(1) ACB abnormal trip

This signal is the output if the generator ACB is tripped by any other means apart from the following: By operation of the ACB CONTROL switch (BCS) or automatic tripping on completion of the load shift control.

(2) Voltage non-established

This signal is the output if the generator does not build up a required voltage within a predetermined time as counted from the instant of its engine's low speed pickup to a specified value (generally 300 rpm).
- Voltage non-establish detect timing (Volt. Non Est. time): 0 to 30 seconds.

(3) Automatic synchronizing failure

This signal is the output if the automatic synchronizing (generator synchronizing and its ACB closure) is not completed within a predetermined time as counted from the instant of an auto synchro start command signal (external or internal).

- Auto synchro failure detect timing (Auto Synchro Fail time): 0 to 240 seconds.

(4) ACB non-closure

This signal is the output when the generator ACB does not close within a predetermined time as counted from the instant of the GAC21 system's giving a CLOSE command to the ACB.

- ACB non-closure detect timing (ACB non-close time): 0 to 10 seconds.

(5) Generator system trouble indication and resetting

① Indication of generator system troubles
Troubles mentioned in (1) to (4) above are individually detected for each generator system, and cause the associated alarm outputs from GAC system to the outside. Usually, these alarms' outputs are used to turn on the visual alarm devices (Push-button switches "ACB TROUBLE RESET (ATR)" indicator lamp color red) provided on each associated generator panel of the main switchboard.

② Resetting of generator system's trouble alarm output
When a generator system trouble alarm output has turned on the visual alarm device mentioned above, pressing this device resets the alarm output and turns off the alarm light in the device. For the voltage non-established alarm output, resetting is possible when the generator's engine speed is below the predetermined speed pickup point or after the generator's voltage has established (Wang et al. 2017) (i.e., pressing the push-button switch does not reset the voltage non-established alarm output if the engine is operating at synchronous speed and the generator is not establishing voltage).

4.1.3 Detection of generators available for automatic start-stop control

The status of the engine, ACB, and control switches are monitored for each generator to determine if the generator can be automatically started or stopped in response to a bus trouble, under the "Number of Diesel Generators on Line control" or by means of an external start signal (e.g., preference trip signal) or an external stop signal (Yang et al. 2010).

(1) Detection of "Ready for Automatic Start" conditions

In general, a generator can be automatically started when all of the following conditions are satisfied:

① The generator engine's "CONTROL POSITION" select switch (COS-L) is set to "REMOTE" position.

② The engine's start interlock conditions are satisfied or the engine is already rotating at speeds above its predetermined speed pickup point.

③ All engine trouble alarm signals are reset and are in normal state.

④ The engine is not under stop control.

⑤ All generator system trouble alarm signals are reset and in normal state.

⑥ The ACB is not closed.

⑦ The associated following control switches are set to the specified positions:

EXAMPLE

The "MODE SELECTION" switch (COS-A): "MANU" – "AUTO" is set to "AUTO" position.

The "SYNCHRO & POWER CONTROL" switch (COS-P): "MANU" – "AUTO" is set to "AUTO" position.

The "CONTROL POSITION" select switch (COS-M) : "MSB" – "ECC" is set to "ECC" position.

The starting order of generators ready for automatic start is determined by the "Stand-by" mode.

(2) Detection of "Ready for Automatic Stop" conditions

In general, a generator can be automatically stopped (removed from the line) when all of the following conditions are satisfied:

① The generator is on line with its ACB closed.

② The associated control switches are set to the specified positions (See VII) of (1) above.

③ The generator is in the parallel operation.

The stopping order of generators ready for automatic stop is determined by the "Automatic Stop Sequence Selection" function. The generators specified as being in the single running order will not stop automatically.

4.2 Bus Monitoring

The GAC Controller monitors the voltage and frequency of the bus to detect bus troubles as mentioned below.

4.2.1 Bus voltage monitoring

This function monitors if the bus voltage is within predetermined limits (BUS VOLT LOW and BUS VOLT HIGH), and outputs the bus low voltage or high voltage signal with a time delay to external device or system when the bus voltage has been below the BUS VOLT LOW setting or above the BUS VOLT HIGH setting. When necessary, these signals may be used to cause automatic starting of a stand-by generator(s).

Voltage settings

- BUS VOLT HIGH: 100 to 125% of rated voltage;
- BUS VOLT LOW: 75 to 100% of rated voltage;
- Time delay (BUS VOLT HIGH TIME and BUS VOLT LOW TIME): 0 to 30 seconds.

4.2.2 Bus frequency monitoring

This function monitors if the bus frequency is within predetermined limits (BUS FREQ LOW and BUS FREQ HIGH), and outputs the bus low frequency or high frequency signal with a time delay to external devices or system when the bus frequency has been below the BUS FREQ LOW setting or above the BUS FREQ HIGH setting. When necessary, these signals may be used to cause automatic starting of a stand-by generator(s).

Frequency settings

- BUS FREQ HIGH: 100.0 to 120.0% of rated frequency;
- BUS FREQ LOW: 80.0 to 100.0% of rated frequency;
- Time delay (BUS FREQ HIGH TIME and BUS FREQ LOW TIME): 0 to 30 seconds.

4.2.3 Two-stage bus monitoring

The above bus voltage and frequency monitoring may be performed in two stages, light trouble (High/Low) and heavy trouble (High/Low), giving the corresponding alarm outputs to external devices or system. When necessary, these alarm signals may be used to cause automatic starting of a stand-by generator(s) (Wang et al. 2017).

4.3 Automatic Generator Start Control

The automatic generator start control is initiated by the detection of a bus trouble or as part of the "Number of Diesel Generators on Line Control", or by the external auto start command signal. This control covers from the automatic starting of the generator engine to the automatic closing of the generator ACB, as appropriate for the particular power generating plant to which the GAC controller is applied (Zhang et al. 2016).

4.3.1 Stand-by generator selection

Stand-by generators are selected by the GAC controller functions. For generators ready for automatic start, this function determines their starting orders as stand-by that responds to automatic start commands. The stand-by select scheme differs depending on the "MODE SELECT" switch (COS-A) arrangement. The following describes an example where five diesel generators are standing by.

(1) One mode select switch for each generator

When mode select switches (one/generator) as in Figure 6 are provided on the main switchboard or engine control console, the stand-by generator's select function checks each generator for "ready for automatic start" conditions and selects the generator's ready for automatic start as 1st stand-by, 2nd stand-by and so on in the order they become ready for automatic start.

Since setting the mode select switch from "MANU" to "AUTO" position is one of the "ready for automatic start" conditions, the order of stand-by generators is usually determined by the temporal order of setting of their select switches from "MANU" to "AUTO" position.

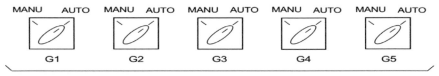

Figure 6. Example of MODE SELECT switch.

SELECTION EXAMPLE

The select switches for G1 and G3 and G5 are set to "AUTO" position as illustrated above. G2 are on line. G1 became ready for automatic start first, G3 second, and G5 third.

① Order of Stand-by Generators

Stand-by: 1st → 2nd → 3rd

Generator: G1   G3   G5

② Moving Up from Lower Order to Higher Order
With the stand-by generator list as in ① above, if G1 is put out of the "ready for automatic start" conditions, then G1 will be removed from the stand-by generator list and the order of the other stand-by generators lower in order than G1 will be moved up as follows:

Stand-by: 1st → 2nd
        | |
Generator: G3   G5

③ Addition of a New Stand-by Generator
With the stand-by generator list as in ① above, if G1 becomes ready for automatic start, then G1 will be added to the bottom of the existing stand-by generator list as follows:
Stand-by: 1st → 2nd → 3rd
       | | |
Generator: G3   G5   G1

(2) Single Mode Select Switch to Cover All Generators

Illustrated Figure 7 is clockwise direction on the select switch, starting from the position to which the select switch is set.

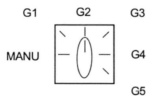

Figure 7. Example of STANDBY SELECT switch.

When a "STANDBY SELECT" switch as illustrated above is provided on the main switchboard or generator control console, the stand-by generator mode select function checks this switch position and each generator for "ready for automatic start" conditions and selects the generators ready for automatic start as 1st stand-by, 2nd stand-by, 3rd stand-by and so on to enable in ascending order of the generator numbers.

SELECTION EXAMPLE

The select switch is set to "G2" position as illustrated above. Generator G1 are on line, and generators G2, G3 and G5 are ready for automatic start.

① Order of Stand-by Generators
Stand-by: 1st → 2nd → 3rd
       | | |
Generator: G2   G3   G5

② Moving Up from Lower Order to Higher Order
With the stand-by generator list as in ① above, if G3 is put out of the "ready for automatic start" conditions, then G3 will be removed from the stand-by

generator 1st and the order of the other stand-by generators lower in order than G3 will be moved up as follows:

Stand-by: 1st → 2nd
          | |
Generator: G2   G5

③ With the stand-by generator list as in ① above, if G3 becomes ready for automatic start, then G3 will be included in the stand-by generator list with its order as determined from the select switch position. The new stand-by generator list will be as follows:

Stand-by: 1st → 2nd → 3rd
          | | |
Generator: G2   G3   G5

4.3.2 Stand-by generator start control schemes

The scheme of stand-by generator start control in response to an automatic start command signal to allow for the selection of the most suitable scheme for individual applications is in the following three kinds. This control is performed by the GAC controller functions.

(1) Single start scheme without followers

In this scheme, only the 1st stand-by generator is automatically started and controlled. If this generator is removed from the stand-by generator list due to a start failure or later control trouble, the stand-by start control is terminated without any further action.

(2) Follower start scheme

If the 1st stand-by generator is removed from the stand-by generator list due to a start failure or later control trouble, the order of the remaining stand-by generators will be moved and stand-by will be started and controlled as 1st stand-by. A failure of this stand-by too will be similarly followed and covered by the next lower order stand-by. This sequence would be continued as long as there is a follower stand-by. The stand-by start control will be terminated when there is no follower stand-by.

(3) Simultaneous start scheme

This control scheme allows the 1st and 2nd stand-by generators (and other stand-by generators if necessary) to simultaneously start. This scheme is divided into two control types: one type of standard arrangement in which the generator, whose voltage is established earlier, is put on line as the 1st stand-by. The other type of request arrangement is the one in which the 1st stand-by and 2nd stand-by generators are sequentially put on line in this order.

4.3.3 Causes of automatic stand-by start and control after automatic start

The generator that is the 1st stand-by on the stand-by generator list is automatically started by the automatic start command signal, which is generated on one of different causes. In the case of follower start scheme, depending on the type of cause, the 2nd stand-by on the list may also be started simultaneously. In the case of simultaneous start scheme, the order of stand-by on the stand-by generator list may not agree with the actual order of stand-by after voltage establishment. Operation control of the started stand-by generator after its voltage establishment differs, depends on the type of cause of automatic stand-by start such as whether it is paralleled to the generator(s) on line or it replaces the latter (Qian et al. 2010).The following describes the automatic start. This control is performed by the GAC controller functions.

(1) Automatic control on bus blackout

The bus blackout signal is produced on abnormal tripping of a generator ACB which results in the following conditions: dead bus trouble signal input from external device and all open generator ACBs.

Two types of automatic control are available for this cause:

① 1st stand-by is started and put on dead bus.
② 1st and 2nd stand-by is started, the 1st stand-by is put on dead bus, and then the 2nd stand-by is synchronized to bus and put on line.

(2) Automatic control on bus trouble

When the bus monitoring is made in two stages of minor and serious troubles, ("Two–stage Bus Monitoring") different types of operation control are available for each level of trouble for the choice. When the bus monitoring is in single stage, the choice may select one of the types of control for serious trouble.

① Automatic control on light Bus trouble
(a) Alarm output only, no automatic stand-by start.
(b) Started stand-by is always synchronized to bus and put on line.
(c) Started stand-by is synchronized to bus and put on line if bus trouble exists at the time of its voltage establishment, otherwise it is kept idling.

② Automatic control on heavy bus trouble
(a) Alarm output only, no automatic stand-by start.
(b) Upon voltage establishment of started stand-by, generator on line is always removed and replaced by the stand-by generator (blackout changeover).
(c) Same as (b) if bus trouble exists at the time of stand-by voltage establishment, otherwise started stand-by is synchronized to bus and put on line.

(d) Same as (b) if bus trouble exists at the time of stand-by voltage establishment, otherwise started stand-by is kept idling.

For types (b), (c), and (d) of automatic control ②, two standbys can be sequentially put on line. In this mode, a serious bus trouble will cause the 1st and 2nd stand-by generators to simultaneously start. After the 1st stand-by generator undergoes specified automatic operation control and the ACB is closed, the 2nd stand-by generator is automatically synchronized to the bus and put on line.

(3) Automatic paralleling to generator(s) on line

In this control scheme, one or more stand-by generators are started on the automatic start command signal, synchronized to the bus, and put on line for parallel operation with the generator(s) on line. The control scheme is in two types, depending on the number of stand-by generators to be put on line.

① Single stand-by synchronized and put on line
In this type of control, 1st stand-by is started on one of the following causes, synchronized to the bus and put on line:
- Abnormal tripping of ACB of one of generators on line (parallel operation).
- Automatic start command under the "Number of Diesel Generators on Line Control" (GAC-PMS-G).
- Automatic start command under the "Large Motor Start Blocking Control" single stand-by start (GAC-PMS-M).

Following causes are detected external to the GAC system:

- Preference trip system operation signal.
- Other signals.

② Two standbys synchronized and put on line
In this type of control, 1st and 2nd stand-bys are started on one of the following stand-by:
- Automatic start command under the "Large Motor Start Blocking Control" double stand-by start (GAC-PMS-M).
- Other causes detected external to the GAC system.

(4) Generator on line automatically replaced by stand-by(s) without blackout period

In this control scheme, one or more stand-by generators are started on the automatic start command signal, synchronized to the bus, and put on line. Then predetermined generator on line is automatically removed after load shift control.

The control scheme is in two types depending on the number of stand-by generators to be put on line.

① Non-blackout changeover to single stand-by
   In this type of control, 1st stand-by is started on an external automatic start command signal. The starting stand-by replaces the predetermined generator on line.

② Non-blackout changeover to two stand-by
   In this type of control, 1st and 2nd standbys are started on an external automatic start command signal. The 1st stand-by is synchronized and put on line. Then 2nd stand-by is synchronized and put on line. With the two standbys on line, the predetermined generator on line is removed after load shift control.

4.4 Automatic Generator Stop

Automatic generator stop control is performed as part of the "Number of Diesel Generators on Line Control" (GAC-PMS-G control function).

The automatic stop command signal by the GAC-PMS-G function causes a sequence of automatic operations from the load-shifting, generator ACB tripping and to the stopping of the generator engine. This control is performed by the GAC controller functions.

4.4.1 Automatic stop sequence selection

The stopping order of generators ready for automatic stop is determined by one of two methods, depending on the "MODE SELECT" switch (COS-A) arrangement (same switch as used for "Stand-by Generator Selection").

(1) One mode select switch for each generator

When mode select switches (one/generator), as illustrated above, are provided on the main switchboard or engine control console, the stopping sequence select function checks each generator on line for "ready for automatic stop" conditions and selects the generators ready for automatic stop as 1st stopping, 2nd stopping, 3rd stopping and so on, in the reverse sequence of their becoming ready for automatic stop.

SELECTION EXAMPLE

The mode select switches for G2, G3 and G5 are set to "AUTO" positions as in Figure 6. G2, G3, and G5 were put on line in this order and all of them were ready for automatic stop. The generator specified as being in the last stopping order would not stop automatically.

① Stopping Orders
   The stopping generator list in this example will be as follows:

Stopping Order: 1st → 2nd

Generator: G5   G3

② Addition of a New Generator on Line

With the stopping generator list as in ① above, if G1 is newly put on line and is ready for automatic stop, then G1 will be added on top of the existing stopping generator list as follows:

Stopping Order: 1st → 2nd → 3rd

Generator: G1   G5   G3

③ Change of Stopping Orders

With the stopping generator list as in ① above, if G3 trips abnormally (assuming that the 1st stand-by generator is G1), G3 is removed from the stand-by generator list and the order of remaining generators is raised as follows:

Stopping Order: 1st

Generator: G5

At this time, if the ACB of G1 in place of G3 is closed automatically and synchronously, the order of stopping generators is as follows (See also ② above):

Stopping Order: 1st → 2nd

Generator: G1   G5

(2) Single standby select switch to cover all generators

When a "STANDBY SELECT" switch as in Figure 7 above is provided on the main switchboard or generator control console, the stop sequence select function checks this switch position and each generator on line to be ready for automatic stop conditions and selects the generators ready for automatic stop as 1st stopping, 2nd stopping, 3rd stopping and so on enabled in descending order of generator numbers. (Illustrated Figure 7 is counter-clockwise direction on the select switch) starting from the position to which the switch is set.

SELECTION EXAMPLE

The stand-by select switch is set to "G2" position as in Figure 7 above. Generators G1, G2, and G5 are on line and ready for automatic stop. In this case, the stopping generator list will be as follows:

The generator specified as being the last stopping order will not stop automatically.

Stopping Order: 1st → 2nd

Generator: G1 G5

4.5 Automatic Load Sharing Control

The bus frequency and kW load on each generator on line are monitored, based on which the governor control signals are outputted to the associated generator engine governors to maintain the bus frequency at the rated value and to properly divide the total load among the generators on line according to the predetermined load sharing mode (Wang et al. 2014).

The "SYNCHRO & POWER CONTROL" switch (COS-P) does not come standard with the system. When the "SYNCHRO & POWER CONTROL" switch is added and is set to the "AUTO" position. The automatic load sharing function will be alive, irrespective of control mode selector switch (COS-A) setting to "AUTO" or "MAMU" positions.

This control is performed by the GAC controller functions.

- Load sharing control accuracy: 5% of rated generator kW output (in reference to mean loading percentage).
- Frequency control accuracy: 0.3 Hz.
- Maximum LOAD LIMIT setting: 50 to 100% of rated generator kW output.
- SHARING RATIO setting: 1.0 to 5.0
- Governor Characteristic setting: 2.0 to 16.0 (sec/Hz).

4.5.1 Types of load sharing control

(1) Proportional Load Sharing Control

The total kW load is divided among the generators on line so that their loading percentage values will be proportional to their SHARING RATIO settings, where the loading percentage of a generator is defined as follows:

$$\textbf{LOADING PERCENTAGE} = \frac{\textbf{kW Load on Generator}}{\textbf{Rated kW Load of Generator}} \times 100\%$$

(2) Optimum Load Sharing Control

In the optimum load sharing control, the SHARING RATIO of each generator should be set to an equal value. See Section 4.2 "Optimum Load Sharing Control (GAC-PMS-TG/SG)" for full information (Wang et al. 2015).

(3) Load Shift Control

If a load shift start signal is given by the automatic start-stop control function or from an external source when more than one generator are on line, load on a predetermined generator is shifted to the other generator(s) on line. When load on this generator is reduced to nearly zero (5% or less of its rated kW output), its ACB is automatically tripped. That is, the target kW load share of a generator under load shift control is 0 kW. If it is expected that load on the other generator(s) will exceed the Load Sharing Cancel setting (LOAD LIMIT) as a result of load shift control, the load shift signal will not cause the load shift operation.

4.5.2 Automatic frequency control

Regardless of the number of generators on line, this control function maintains the bus frequency at the rated value, controlling the governor motor of each generator engine. In case of single operation, the GAC Controller provides the control including governor motor control, providing that the stand-by control "MODE SELECT" switch (COS-A) has been set to "AUTO" position.

The "SYNCHRO & POWER CONTROL" switch (COS-P) is provided, the automatic frequency control does function if this switch is set to "AUTO" or "MANU" position which takes priority, irrespective of control mode selector switch (COS-A) setting to "AUTO" or "MANU" positions.

4.5.3 Generator engine governor control

Actual kW load on each generator on line is monitored and compared with the target kW load share mentioned in Section 3.5.1 to compute:

- **Power Deviation = Target kW Share - Actual kW Load**.
  The bus frequency is also monitored to compute:
- **Frequency Deviation = Rated Frequency - Actual Bus Frequency**.

According to whether these deviations are positive or negative in value, one of two governor control signals is selected as follows:

- Positive Deviation … Governor Raise (UP) signal;
- Negative Deviation … Governor Lower (DOWN) signal.

These control signals are produced with respect to power deviation and frequency deviation.

According to the magnitude of the above deviations and based on the governor characteristic (GOV CHARACT) [sec/Hz] setting:

The pulse duty ratio of these governor control signals is determined. The governor control signal has been detected as output from the GAC controller to the generator that is now operating at the frequency of once every 5 seconds and the maximum governor drive time of 2.4 seconds (8 seconds/Hz). The governor control signal is the output based on the resultant value of power deviation and frequency deviation.

Governor control function is comprised of a governor control for load share and a governor control for frequency adjustment, a sum per time is the output. During ACB closed, constant frequency control; During ACB opened and automatic synchronize control by GAC, generator's revolution speed control for synchronizing, the governor output is the sum of GVp and GVf.

$$GVp\ (sec) = \frac{\text{Power deviation } (KW) \times \text{Governor speed } (sec/Hz) \times [\text{Droop}]\ (\%) \times [\text{Load share control gain}]\ (\%)}{\text{Generator rated power } (KW) \times 10000}$$

$$GVf\ (sec) = \frac{\text{Frequency deviation } (Hz) \times \text{Governor speed } (sec/Hz) \times [\text{Frequency control gain}]\ (\%)}{100}$$

Where, GVp: Governor in quantity for load share control;
GVf: Governor in quantity for frequency control.

4.6 Automatic Synchronous Closing Control Function

To perform the automatic synchronous closing control, use the Digital Synchronizer which is one of the components for the GAC System. This function is used to place a generator on a live bus served by other generator or generators for parallel operation with the latter. This function compares the frequency of a started generator that has built up the voltage with the bus frequency and brings the generator frequency closer to the bus frequency through governor control (Huang and Huang 2007). When the differences in voltage and frequency between the generator and bus are within their predetermined limits, the governor control finishes. If the frequency difference or voltage difference becomes smaller than the predetermined value, the Digital Synchronizer outputs the ACB closing signal to the outside via the GAC controller, so that the ACB is closed when the bus voltage phase matches with the generator voltage phase.

Specifications

- Acceptable frequency difference limit: 0.1 to 0.5 Hz
- Acceptable voltage difference limit: 2 to 10% of bus voltage
- Maximum permissible frequency difference for operation: 8 Hz
- Governor characteristic setting range: 2.0 to 16.0 (sec/Hz)
- Circuit breaker closing time: 0 to 500 ms
- Automatic synchronizing fail time: 0 to 240 sec

When a generator is connected in parallel to BUS, automatic synchronizing device is controlled with smooth control and no-shock, an ACB closing ordered signal is output by measuring phase difference. Automatic synchronizing function is shown as Figure 8.

In Figure 9, allowable frequency band for ACB closed operation is dependent on the setting values of [Frequency Optimum Difference] and [Frequency Optimum Sanction], showed with ////marking, under the mentioned drawings. And in ▨▨ marked band, the stand-by generator's frequency is drawn out because the frequency between BUS and stand-by generator closely comes to stagnation.

The following description of parameter corresponds to synchronizing feeder to bus bar.

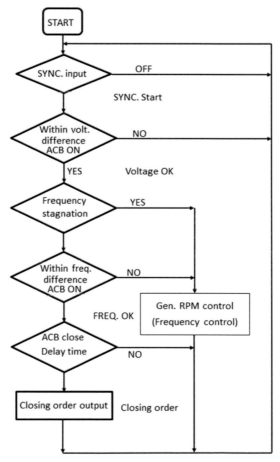

Figure 8. Automatic synchronizing function flow chart.

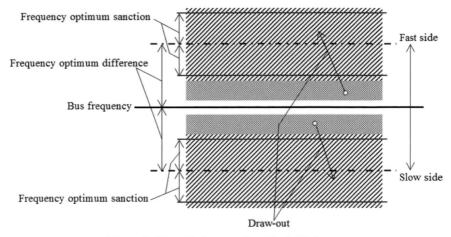

Frequency optimum sanction

Fast side

Frequency optimum difference

Bus frequency

Slow side

Frequency optimum sanction

Draw-out

Figure 9. Allowable frequency band for ACB closing.

Sync. Active: To activate the synchronizing. If it is active, the synchronizing appears and the synchronizer with its frequency and voltage controller is operative.

Max phase angle: The maximum phase angle (positive and negative angle) appears, the closing order for the switching device (breaker) is given.

Max frequency diff.: If the difference of frequency is higher than the set point, then the synchron closing is blocked until the difference of frequency drops below the set point.

Min frequency: Minimum frequency limit for "abnormal bus" check. If the frequency of bus bar is below the minimum frequency limit, then the frequency controller will be deactivated. This means the output to send signals of lower speed will be blocked.

Max frequency: Maximum frequency limit for "abnormal bus" check. If the frequency of bus bar is above the maximum frequency limit, then the frequency controller will be deactivated. That means the output to send signals of higher speed will be blocked.

Max voltage diff.: If the difference in voltages is higher than the set point, the synchron closing will be blocked until the difference in voltages drops below the set point.

Min voltage: Minimum voltage limit for "abnormal bus" check. If the voltage of bus bar is below the minimum voltage limit, then the voltage controller will be deactivated. That means the output to send lower voltage signals will be blocked.

245

Max voltage: Maximum voltage limit for "abnormal bus" check. If the voltage of bus bar is above the maximum voltage limit, then the voltage controller will be deactivated. That means the output to send higher voltage signals will be blocked.

Freq. set pulse time: The pulse time for frequency adjustment during synchronization is set here. The pulse time is defined at 100 percent frequency difference. The pulse time, which the frequency controller calculates after the break time, is modified by the difference between the two frequency inputs. The formula for the pulse is:

Calculation for pulse time:

$$\text{Pulse [sec]} = \text{Parameter}/100 * \Delta F \text{ [\%]}$$

Whereby, Parameter: pulse time set point at 100% frequency difference.

ΔF [%]: frequency difference between feeder and bus bar 1 in percent.

Example:
Parameter = 100 sec means:

If the frequency difference between feeder and bus bar is 100%, a pulse of 100 seconds will be set. If the frequency difference is 1% (0.6 Hz at 60 Hz nominal-rated frequency), then a pulse of 1 sec will be set and so on and so forth.

Freq. set. Breaktime: The breaktime between pulse times for speed adjustment during synchronization is set here. After a breaktime is passed, the frequency controller will calculate the next pulse time to control the pulse event.

Speed push after 10 s: If the difference between the electrical angles of the generator and bus bar does not become zero within 10 seconds, then an impulse of the set time in this parameter will be given.

Volt. Set. Pulse time: The pulse time for voltage adjustment during synchronization is set here. Pulse time is modified by the difference between the voltages inputs.

Volt. Set. Breaktime: The breaktime between pulse times for voltage adjustment during synchronization is set here.

Freq. higher event: Use this event to activate a digital output in order to increase the frequency during synchronization.

Freq. lower event: Use this event to activate a digital output in order to lower the frequency during synchronization.

Volt. Higher event: Use this event to activate a digital output in order to increase the voltage during synchronization.

Volt. Lower event: Use this event to activate a digital output in order to lower the voltage during synchronization.

Closing direction: The main breaker closing command will be given only when the engine speed is higher (UP), lower (DOWN) or in both directions.

CB closing delay: The main breaker closing delay is the mechanical delay time caused by the closing time of contactors, relays, and the main breaker itself. This delay time can reduce the time period of "breaker synchron on command".

Live/dead cond. check: ON/OFF-switch for dead condition check. If the synchronizing unit recognizes one side (feeder or bus bar) without voltage (dead system), then MB sync. ON will be activated without checking the synchronization.

Min. voltage level: If the live/dead condition check is active and the voltage of one side (feeder or bus bar) is below this voltage limit, the synchronizing unit recognizes dead condition.

CB sync. ON: Use this event to send the main breaker close signal. This event will be set if the pointer of the synchronization is within the synchron tolerable limit and the synchronization unit is not blocked.

Abnormal bus: Use this event to indicate bus abnormal situation. If during synchronization the main conditions for voltage and frequency are not fulfilled for the bus bar, then bus abnormal event will be set and the corresponding controller for voltage and frequency will be blocked.

The following diagram is shown as Figure 10, the conditions have to be fulfilled during synchronizing to set the "CB synchron event".

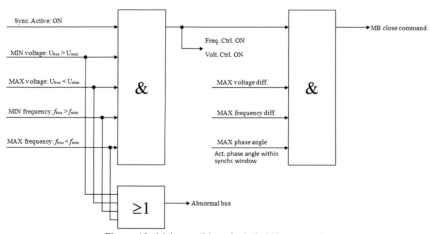

Figure 10. Main conditions for MB ON command.

The closing direction philosophy diagram is shown as Figure 11. If the closing direction "up" is selected and if the feeder frequency is higher than the bus bar frequency, then the main breaker synchron close command will be given.

If closing direction "down" is selected and if the feeder frequency is lower than the bus bar frequency, then the main breaker synchron close command will be given. Its diagram is shown as Figure 12.

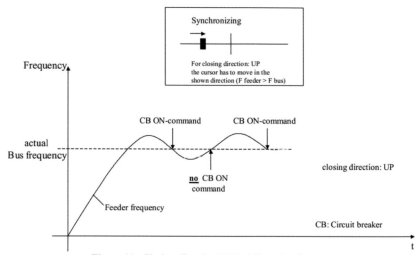

Figure 11. Closing direction "UP" philosophy diagram.

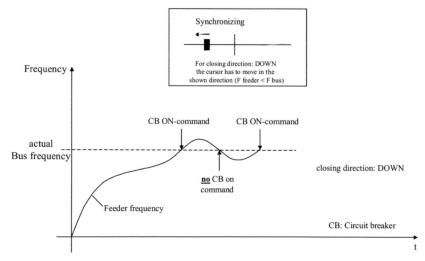

Figure 12. Closing direction "DOWN" philosophy diagram.

In Figure 8 to Figure 12, ACB closing direction is carried out from FAST side, however a particular order of closing from slow side or both FAST/SLOW are available at only primary stage. ACB closing order signal is output in the range from the point GEN ACB DELAY TIME for the time pre-sett by [ACBCLOSE SIGNAL HOLD ON TIME] during SYNC input signal "ON". This function is operated during SYNC input signal "NO". Accordingly, after the ACB closing operation having finished, the SYNC input signal should be "OFF" in electric writing circuits, is shown as Figure 13.

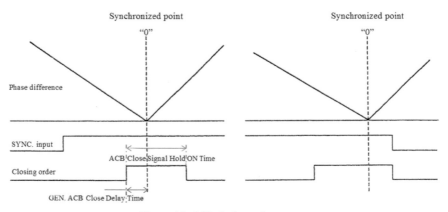

Figure 13. ACB closing order output.

4.6.1 SYNCHRO DETECT function

This function compares the generator voltage signal with the bus voltage signal, both inputted from the external potential transformers, to check if the following conditions are satisfied:

Voltage difference: Difference between the bus and generator voltages is within the acceptable voltage difference limit (normally set to 3%).

Frequency difference: Difference between the bus and generator frequencies is within the acceptable frequency difference limit (normally set to 0.3 Hz).

If these conditions are satisfied, then the Digital Synchronizer function outputs an ACB close command signal to the external generator's ACB control circuit within a certain length of time (ACB closing time, normally set to 75 ms) prior to the instant when the generator voltage comes exactly in phase with the bus voltage.

4.6.2 SPEED MATCHER function

This function compares the generator voltage with the bus voltage, both inputted from the external potential transformers, to detect the difference in frequency between them, according to which one of the following governor control signals is produced:

Governor Raise (UP): Produced when the generator frequency is lower than the bus frequency.

Governor Lower (DOWN): Produced when the generator frequency is higher than the bus frequency.

These governor control signals are pulse signals of interval of maximum 5 seconds and governor driving time of 2.4 seconds (8sec/Hz) by constant pulse repetition period, and the pulse duty ratio varies according to the magnitude of frequency difference and based on the GOV CHARACT (governor characteristic) setting. The governor control signals are outputted from Digital Synchronizer through the GAC controller to the generator engine governor motor.

The governor pulse breadth to be calculated from the frequency deviation, the set of [GOVERNOR SPEED], and the set of [FREQUENCY CONTROL GAIN] is specified as follows and the set of [FREQUENCY CONTROL DELAY TIME] determine the governor output signal.

$$\text{Governor pulse breadth (sec)} = \frac{\text{Frequency deviation (Hz)} \times \text{Governor speed (sec/Hz)} \times \text{Frequency control gain (\%)}}{100}$$

4.7 Generator Low-Load Detection Function

This function detects the low-load operation state of a generator and outputs the detected signal to an external monitor as the warning signal. If the measured power value of a generator during its operation decreases to and remains below the setting value for low load, the signal is the output.

- Low load setting range: 0 to 100% (Rated output)
- Low load time setting range: 0 to 240 seconds

4.8 Warning Indicator Lamp Signal Output

The display program provides generator-related warning output and common in the operation pattern, and is shown as Figure 14.

When the power source of 24 V DC is turned ON, the output signal and control signal are stopped for 2 seconds. For the group warning signal*1, if a new error occurs during the output of the error signal, the system is restored to the normal state (for 2 seconds) and then the error signal is output again.

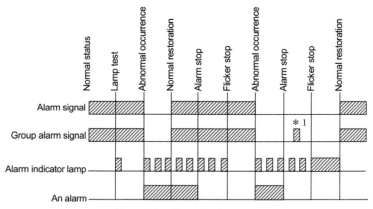

Figure 14. Alarm signal pattern.

5. GAC-PMS Control Functions

Under strict demand for fuel economy and environmental protection, energy conservation, emission reduction, and cost control are the inevitable choice of coping with the shortage of resources and environmental carrying capacity. Modern marine power stations include diesel generator (DG), ship shaft generator (SG), exhaust gas turbine generator (PT-G), and steam turbine generator (ST-G). In order to realize energy saving and emission reduction of ships, their economic benefits and combination methods acquires importance in the shipping field (Hu et al. 2015), as shown in Figure 15 and Figure 16. The optimal load distribution between SG, TG, and DG is the main content of energy saving optimization management of ship power station.

5.1 Number of Diesel Generators on Line Control (GAC-PMS-G)

This function controls the number of diesel generators on line in response to changes in the ship's power demand. This function is intended to ensure a higher degree of "continuity of service" through minimizing operations of the preference tripping system and generator overload protection and also to operate the diesel generators within their high fuel-efficiency region, as shown in Figure 17.

Settings for GAC-PMS-G

- Start request point, K1 (START REQ DG-H): 50 to 100%.
- K_1 timer (START REQ TIME): 0 to 30 seconds.
- Stop request point, K2 (STOP REQ): 50 to 100%.
- K_2 timer (STOP REQ TIME): 0 to 2400 seconds.

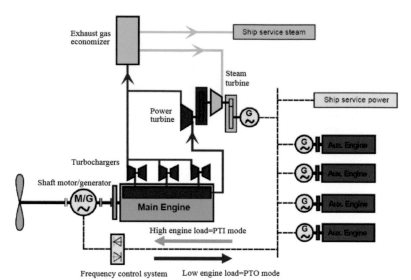

Figure 15. Integrated energy-saving system of advanced marine engine.

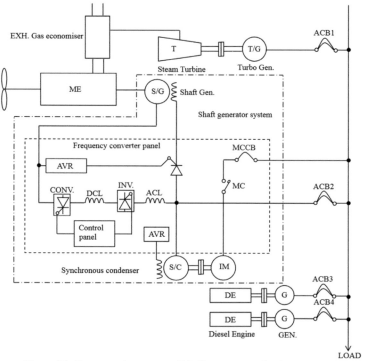

Figure 16. Energy saving system of shaft generator and turbo generator.

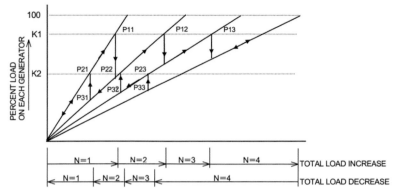

Figure 17. Diagram to show a number of diesel generators on line control.

K₁ and K₂ are mathematical symbols used for explanation purposes only. In setting them, K_1 should be greater in value than K_2.

(1) Start Request (Automatic Start)

In Figure 17, assume the total kW load is increasing. When the loading percentage of the diesel generator or generators on line reaches K_1 (%) and this condition exists for the time length set by the K_1 timer, an automatic start command signal is produced. In response to this signal, the stand-by generator is automatically started, synchronized and put on line for parallel operation with the generator(s) already on line.

This start request point K_1 (%) is mathematically expressed as:

$$K_1 = \frac{\sum_{n=1}^{N} P_{Ln}}{\sum_{n=1}^{N} P_{Rn}} \times 100\%$$

Where P_{Ln}: load on generator all generators in kW;
P_{Rn}: rated output of all generators in kW;
N: number of generators on line.

Where N: number of generators on line.
P_{1N}: (= K1) stand-by start request point (%).
P_{2N}: (= K2) stop request point in expected loading percentage after generator removal.
P_{3N}: actual loading percentage before generator removal at P2N point.

(2) Stop Request (Automatic Stop)

In Figure 17, if the total kW load decreases to such a degree that the removal of the 1st stopping generator will not result in loading percentage more than K2 (%) for each diesel generator remaining on line, and this condition exists for the time length set by the K2 timer, an automatic stop command signal is produced. In response to this signal, load on the 1st stopping generator is shifted to the other generator(s) on line, then the ACB of the first stopping generator is tripped and its engine is automatically stopped (after a predetermined length of engine idling time (ENG IDLING TIME) when necessary).

The stop request point K2 (%) is mathematically expressed as:

$$K_2 = \frac{\sum\limits_{n=1}^{N} P_{Ln}}{\sum\limits_{n=1}^{N} P_n - P_R} \times 100\%$$

Where P_R: rated output of first stopping generator in kW.

5.2 Automatic Load Sharing Control Function

The power plant of the container ship has five electric power-generating sets: four sets of diesel engine driven generator and one set of turbine driven generator. Usually when the ship is at sea, one or two sets of the diesel generator is operated. At the time of calling/leaving ports, and cargo loading/unloading, two or more sets of the generators are put into parallel operation for the power demand (Hao et al. 2014). The diesel generators may also be set as standbys so that in the event of a feeding trouble, the stand-by generator will be automatically started to replace the generator in trouble.

Automatic load sharing control has three kinds of control methods which are described in Table 1. They are chosen by OVF (Over Flow) signal input, where a shared electric power value is decided under running generators by comparing A+B with total powers, A($\sum$High_Limit) shows a generator's high limited total power that is indicated to suggest over flow control method, B($\sum$Low_Limit) shows a generator's low limited total power that is indicated to suggest remaining proportional control method. Automatic load sharing control flow chart is shown as Figure 18.

In case of A+B > total powers, a generator that is ordered to run at over flow control method is carried out at generator's high limited power, the other generator that is ordered to run at proportional control method is carried out proportional

Table 1. Automatic load sharing control list.

Over flow input	Input order	System load condition	Target load to control	Control method
OFF	Proportional control	$\sum$(OPT)High_Limit+ $\sum$(PRO)Low_Limit < Total load	$$\dfrac{\left(\text{Total laod}-\sum(\text{OPT})\text{High}\_\text{Limit}\right)\times\text{TYP}}{\sum(\text{PRO})\text{TYP}}$$	Proportional running*1
		$\sum$(OPT)High_Limit+ $\sum$(PRO)Low_Limit > Total load	$$\dfrac{\left(\text{Low}\_\text{Limit point}\right)(\%)\times\text{TYP}}{100}$$	Low load limit running*2
ON	Over flow control	$\sum$(OPT)High_Limit+ $\sum$(PRO)Low_Limit < Total load	$$\dfrac{\left(\text{High}\_\text{Limit point}\right)(\%)\times\text{TYP}}{100}$$	Load limit running*3
		$\sum$(OPT)High_Limit+ $\sum$(PRO)Low_Limit > Total load	$$\dfrac{\left(\text{Total laod}-\sum(\text{OPT})\text{High}\_\text{Limit}\right)\times\text{TYP}}{\sum(\text{PRO})\text{TYP}}$$	Proportional running*4
		$\sum$Low_Limit > Total load	$$\dfrac{\text{Total load}\times\text{TYP}}{\sum(\text{PRO})\text{TYP}}$$	Proportional running*5
		$\sum$High_Limit < Total load		
		In case of proportional control order for all engines		

Notes:

Total load: Generator's total load power (KW); TYP: Generator's rated capacity (KW);

High_Limit: Generator's high load limit power (KW); OPT: Generator was ordered Over Flow control;

Low_Limit: Generator's low load limit power (KW); PRO: Generator was Proportional control;

$\sum$(OPT)High_Limit: Generator's high load limit total power (KW) for generator that was ordered Over Flow control;

$\sum$(PRO)Low_Limit: Generator's low load limit total power (KW) for generator that was ordered Proportional control.

shared control method of remaining power. Generator's high limited power running at keeping under calculation value is:

$$\frac{[\text{rated power}]\ (KW)\ \times\ [\text{high limited point}]\ (\%)}{100}$$

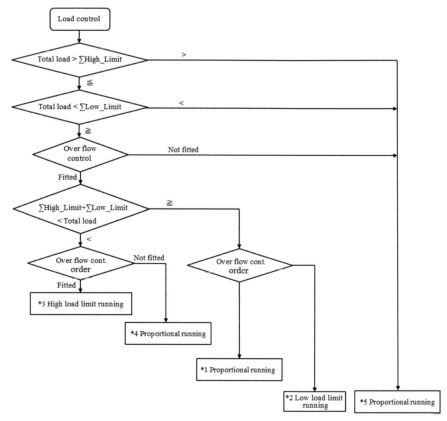

Figure 18. Automatic load sharing control flow chart.

In case of A+B < total powers, a generator that is ordered to run at proportional control method is carried out generator's low limited power running, the other generator that is ordered to run at over flow control method is carried out to run at proportional shared control method of remaining power. Generator's low limited power running at keeping under calculation value is:

$$\frac{[\text{rated power}](\text{KW}) \times [\text{low limited point}] \, (\%)}{100}$$

Automatic load sharing control example is shown as Figure 19, Table 1, and Table 2.

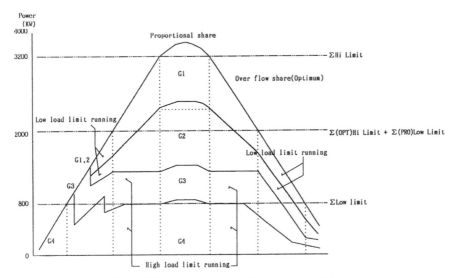

Figure 19. Automatic load sharing control principle.

Table 2. Automatic load sharing control conditions.

Generator No.	Order	Rated capacity (KW)	High load limit (%)	Low load limit (%)
G1	Proportional	1000 KW	80%	20%
G2	Proportional	1000 KW	80%	20%
G3	Over flow	1000 KW	80%	20%
G4	Over flow	1000 KW	80%	20%

5.3 Optimum Load Sharing Control (GAC-PMS-TG/SG)

This control function is shown as Figure 20 and intended to operate the exhaust-gas turbo-generator (TG) or shaft-driven generator (SG) at its maximum efficiency, thereby reducing the total operation hours and maintenance costs of the diesel generator or generators (DG) that back up TG or SG. Figure 20 shows an example of optimum load sharing control that covers TG and DG (Hu et al. 2015, Hao et al. 2014).

When TG or SG, and DG are on line, this function controls them so that TG or SG will be optimally loaded to the conventional rated output value, which is set by means of an external digital switch, and DG will never be loaded below a predetermined low load limit. The optimum load sharing control as mentioned above is achieved through monitoring the load on each generator on line, computing the load to be shared by each generator on line, and controlling the governor motor of each generator engine according to the computed load share.

Settings for GAC-PMS-TG/SG

- TG/DG Conventional rated output for control purposes: 0 to 3000 kW (10 kW minimum setting unit);
- Stand-by start request point: 80 to 100% (50 to 100%: available on request);
- DG low load limit: 0 to 50%;
- TG/SG low load limit: 0 to 50%.

The total kW load P_L, is known as the sum of kW loads on generators on line. The following describes the operation under GAC-PMS-TG for each P_L zone specified in Figure 20.

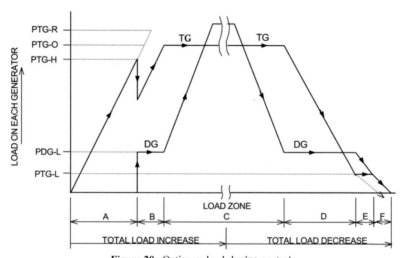

Figure 20. Optimum load sharing control.

(1) Load Zone A

TG is in single operation and load on it increases. When load on TG reaches PTG-H and this condition exists for a predetermined length of time (the K_1 timer setting mentioned in 4.1), an automatic start command signal is given to a stand-by generator (DG). In response to this signal, DG is automatically started, synchronized and put on line for parallel operation with TG.

(2) Load Zone B

DG is loaded to its low load limit PDG-L constant, while the rest of load is shared-by TG. For control in this load zone, the target load share for each generator is computed as follows:
DG: DG rated output × DG low load limit setting (PDG-L) (kW).
TG: PL – PDG-L (kW).

(3) Load Zone C

If load on TG reaches PTG-O (conventional rated output of TG in kW set by means of an external digital switch) with a further increase of P_L, DG low load limit control is canceled. Instead, TG is loaded to PTG-O constant and DG shares the rest of load. For control in this load zone, the target load share for each generator is computed as follows:
DG: PL – PTG-O (kW), provided P_L > (PTG-O + PDG-L).
TG: PTG-O (kW).

That is, changes of P_L above P_L = PTG-O + PDG-L are covered by DG. Where more than one DG are available as standbys, the GAC-PMS-G function controls the number of diesel generators on line to cover changes in P_L.

(4) Load Zone D

If a decrease of the total load P_L results in a reduction of load on DG to its low load limit PDG-L, TG optimum load sharing control is canceled. Instead, DG is loaded to PDG-L constant and TG shares the rest of load. Therefore, the target load share of each generator is as mentioned in Load Zone B.

(5) Load Zone E

If the load on TG is reduced to its low load limit PTG-L as a result of a further decrease of P_L, DG low load limit control is canceled. Instead, TG is loaded to PTG-L constant and DG shares the rest of load. This is to prevent the reverse power flow into TG. For control in this load zone, the target load share for each generator is computed as follows:
DG: P_L – PTG-L (kW).
TG: TG rated output × TG low load limit setting = PTG-L (kW).

(6) Load Zone F

If the load on DG in percent of its rated output becomes equal to the TG low load limit setting (set in % of PTG-L) with a still further decrease of P_L, TG low load limit control is canceled. Instead, TG and DG are controlled in a proportional load-sharing mode with PTG-R being the rated output capacity of TG. This is to prevent the reverse power flow into DG.

The set point of PTG-H is expressed in percentage with respect to the TG conventional output PTG-O (kW). In this example of optimum load sharing control, the last diesel generator remaining on line is intentionally not removed. It is also possible to design the control scheme to remove this diesel generator.

5.4 Large Motor Start Blocking Control (GAC-PMS-M)

The starting current of a large auxiliary's motor, such as the side thruster or ballast pump, can cause serious service disturbances leading to a blackout of the electric power system. This function is intended to prevent such troubles by securing, when necessary, the required power for starting of a large motor, is shown as Figure 21.

Specifications

- Setting for required starting power: $*$ 0 to 2000 kW.
- Number of motors under control: 16 motors maximum.
- Setting for required starting confirmation time: $*$ 0 to 60 seconds [standard setting time: 15 seconds]. Closing operation of the stand-by generator ACB will output the motor start enable signal.

5.4.1 Detection of start enable signal

For each large motor under this control, the required starting power P_{Mn} in kW is set. Large motor starter circuit is shown as Figure 21. Normally, each P_{Mn} value is compared with the total surplus power P_S (kW) as computed from the loading conditions of the generators on line, to give the start enable signal to the starter circuit (See the above diagram) of each large motor under this control.

(1) When $P_S \geqq P_{Mn}$:

This means that there is sufficient surplus power to allow for starting that motor and the associated start enable signal is "ON". Therefore, pressing the START push-button for that motor starts it.

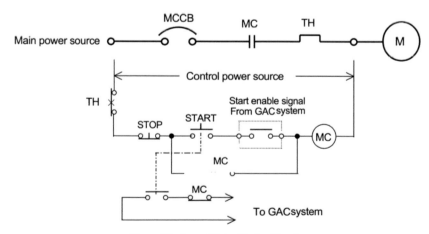

Figure 21. Large Motor Starter Circuit.

(2) When $P_s < P_{Mn}$:

The available surplus power is not sufficient to allow for starting that motor and the associated start enable signal is "OFF". Pressing the START push-button does not start the motor (start blocking).

Instead, it causes one or more stand-by generators to automatically start and put on line, after which the START push-button should be pressed again to start the motor.

5.4.2 Detection of number of generators to be put on line

In the case of Ps < P_{Mn}, pressing the START push-button applies a contact closure signal to the GAC system. In response to this, how many stand-by generators, one or two, have to be started is determined and the corresponding numbers of automatic stand-by start command signals are produced. As a result, the stand-by generator(s) is automatically started and put on line.

The number of stand-by generators to be started is determined as follows.

The expected total surplus power P_{SS} (kW) when the 1st stand-by generator whose rated output is PG1 (kW) is put on line is computed, i.e., $P_{SS} = P_s + PG1$. The start-blocked motor's P_{Mn} value is then compared with P_{SS}, and

(1) If $P_{SS} \geq P_{Mn}$:

An automatic start command signal is produced to start and put on stand-by generator on line (single stand-by start).

(2) If $P_{SS} < P_{Mn}$:

Two automatic start command signals are outputted in parallel from the GAC21 system to start and put two stand-by generators on line (double stand-by start).

6. Communication Function with Monitor Systems

The human interface module of the GAC system outputs warning signals to the Generator Monitor System via the multiplex transmission. All the warning signals are outputted as NOR-CLOSE and the following ones are standard:

(1) Common output signals

- MSB PREF TRIP
- MSB BUS VOLT HIGH
- PREF TRIP/EMERG STOP SOURCE ABNORMAL
- MSB BUS VOLT LOW
- MSB 440V CIRCUIT LOW INSULATION
- MSB BUS FREQ HIGH

- MSB BUS FREQ LOW
- MSB DC24V SOURCE FAIL

(2) Output signals for each generating plant

- G/E OVER SPEED TRIP
- GEN OVER CURRENT
- G/E LO INLET LOW P TRIP
- GEN ACB NON CLOSE
- G/E CFW OUTLET HIGH TEMP TRIP
- GEN ACB ABNOR TRIP
- G/E START FAIL
- GEN ACB SYSTEM FAIL
- G/E STOP SOLENOID ABNORMAL
- G/E LOAD LIGHT
- G/E REPOSE SIG
- G/E RUN HOUR (ACB ON)

7. BUS Monitoring Function

BUS monitoring function is given effect to BUS (R-S phase) voltage and frequency.

When BUS voltage exceedingly fluctuates over the based voltage, the four alarms mentioned below are provided, as shown in Figure 22. Simultaneously, the BUS monitor HIGH VOLT lamp and the BUS monitor LOW VOLT lamp on the operation panel also light-up. It is available to be pre-set separately at an alarm level and a delay time for an alarm output.

The frequency usually monitors the BUS frequency. When the frequency exceeds each alarm sets, the high alarm (H) and the low alarm (L) operate and simultaneously a high frequency lamp or the low lamp on the operation panel lights-up.

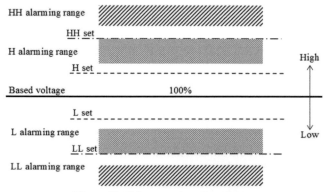

Figure 22. Voltage monitors four alarms.

It is able to set the level in the range of 0–130% of [based voltage], the range of 0–130% of [based frequency], the detecting time in the range of 0–9999 seconds.

8. Conclusions

Remote semi-automatic control and monitoring functions, GAC-UMS control functions, and GAC-PMS control functions have been introduced in this work. The principle of GAC system has been analyzed. The procedure of operation of optimum load sharing control method has been explained in this paper. Merchant ships will benefit greatly from energy conservation, emission reduction, and cost controlling GAC system. Current development efforts must focus on advanced GAC functions and artificial intelligence.

References

Changbin Hu, Xin Wang, Shanna Luo, Zhengxi Li, Xu Yang and Li Sun. (2015). Energy optimization coordination control of multi-sources in microgrid. Proceedings of the CSEE, vol. 5, Suppl., pp. 36–43.

Guichen Zhang and Jie Ma. (2011). Speed-torque hybrid control with self-tuning for marine power station. Journal of Shenyang University of Technology, vol. 33, No. 4, pp. 434–438.

Guichen Zhang. (2013). Modern Marine Power System, Dalian Maritime University Press. China, pp. 180–184.

Jialin Wang, Li Xia, Zhengguo Wu and Xuanfang Yang. (2012). Optimal phasor measurement units placement considering multi-objectives in shipboard power system. High Voltage Engineering, vol. 38, No. 5, pp. 1267–1273

Jinfan Jiang. (2005). Ship Power Station and Automation. Dalian Maritime University, Press. China, pp. 137–142.

Manlie Huang and Changhong Huang. (2007). Research on double-pulse H-infinity speed governor for diesel eingine of ship power station. Control Theory & Application, vol. 24, No. 2, pp. 283–288.

Mei Qian, Zhengguo Wu and Jianghui Han. (2010). Real-time performance analysis of integrated power system monitoring and control network in vessel. Power System Protection and Control, vol. 38, No. 15, pp. 38–46.

MV COSCO BELGIUM. (2013a). Main Generator Instruction. Shanghai. Ocean Shipping Co., Ltd., China.

MV COSCO BELGIUM. (2013b). High Voltage Main Switchboard Instruction. Shanghai. Ocean Shipping Co., Ltd., China.

MV COSCO XIAMEN. (2004). Low Voltage Main Switchboard Instruction. Shanghai. Ocean Shipping Co., Ltd., China.

Shilong Xue. (2011). Ship Power System and Automatic Control. Publishing House of Electronics Industry, Beijing. China, pp. 176–186.

Wei Zhang, Weifeng Shi and Hongqian Hu. (2016). Research on agent based reconfiguration and its optimization for shipboard zonal power systems. Power System Protection and Control, vol. 44, No. 4, pp. 9–15.

Wu Yang, Hanhong Jiang, Chaoliang Zhang and Zhongyuan Hou. (2010). Shipboard Power System Monitoring-Oriented Hybrid Network Technology. Power System Technology, vol. 34, No. 4, pp. 194–198.

Xinzhi Wang, Li Xia and Chao Zhang. (2014). Transient voltage stability analysis for ship power system considering load dynamic model. Journal of Central South University (Science and Technology), vol. 45, No. 7, pp. 2231–2236.

Yanwei Wang, Fan Liu, Tao Ding, Jibo Sun, Zhaohong Bie and Yinguo Yang. (2017). Dynamic active set algorithm for real-time energy-reserve optimal dispatch in multi-zonal power system. Journal of Xi'an Jiao Tong University, vol. 51, No. 4, pp. 45–52.

Yi Wang, Ning Zhang and Chongqiong Kang. (2015). Review and prospect of optimal planning and operation of energy hub in energy internet. Proceedings of the CSEE, vol. 35, No. 22, pp. 5669–5681.

Yuchen Hao, Xiaobo Dou, Zaijun Wu, Minqiang Hu, Chunjun Sun, Tao Li and Bo Zhao. (2014). Hierarchical and distributed optimization of energy management for microgrid. Electric Power Automation Equipment, vol. 34, No. 1, pp. 154–162.

Index